D0929265

# CARBOHYDRATE BUILDING BLOCKS

Mikael Bols

Associate Professor
Aarhus University

A Wiley-Interscience Publication

John Wiley & Sons, Inc.

New York / Chichester / Brisbane / Toronto / Singapore

This text is printed on acid-free paper.



***Library of Congress Cataloging in Publication Data:***

Bols, Mikael
Carbohydrate building blocks / Mikael Bols.
p. cm.
Includes indexes.
ISBN 0-471-13339-6 (cloth : alk, paper)
1. Carbohydrates. 2. Stereoisomers—Synthesis. I. Title.
[DNLM: 1. Carbohydrates—chemistry—handbooks. 2. Stereoisomers—handbooks. 3. Chemistry, Organic—handbooks. QU 39 B693c 1994]
QD321.B73 1996
547.7′80459—dc20
DNLM/DLC
for Library of Congress 95-16543
CIP

Printed in the United States of America

10 9 8 7 6 5 4 3 2 1

To My Wife Kari

# CONTENTS

# FOREWORD

Many routes have been devised for the construction of complex asymmetric structures in enantiomerically pure form. One of the most fruitful of these takes advantage of readily available optically pure starting materials of *known* absolute configuration. The readily available carbohydrates and their numerous, easily obtained derivatives make an unequaled source of such enantiomerically pure starting materials. Many examples have been recorded of imaginative transformations of carbohydrates into impressively diverse targets that often give no hint of their carbohydrate parentage. Unfortunately, recognition of potential carbohydrate starting materials is greatly hampered by the arcane nature of carbohydrate nomenclature, at least for the uninitiated: It is not immediately clear to molecular architects what the structure for what is known to *Chemical Abstracts* as 1,5-anhydro-2-deoxy-D-*arabino*-hex-1-enitol could possibly be, or that it is a potentially extremely useful starting material that is available from D-glucose in three simple steps with an approximately 50% overall yield.

Mikael Bols has done a superb and much needed job of remedying these two difficulties. I am sure that there will be great interest in his original and effective presentation of the map to the largely secret treasures which have long been hidden in the carbohydrate field.

Gilbert Stork

Eugene Higgins Professor Emeritus of Chemistry
Columbia University

# PREFACE

The idea of using carbohydrates as synthons for the synthesis of chiral compounds is certainly not new. Even in Reichstein's 1934 synthesis of L-ascorbic acid, carbohydrates were used as starting material. Since that legendary work carbohydrate synthons have been used many times. Chemists were made particularly aware of the advantages of tapping the carbohydrate chiral pool by Stephen Hanessian's excellent book, *The Chiron Approach.* However, his book focused mainly on the later stages of natural products synthesis and did not serve to increase substantially the synthetic organic chemist's arsenal of starting materials. So despite the greater awareness of carbohydrate synthons in recent years, the full potential of the carbohydrate chiral pool is still little exploited. The goal of this book is to unravel secrets from carbohydrate chemistry that will be highly useful for the synthetic chemist and that are extremely difficult to learn by other means.

I am very grateful to Professor Gilbert Stork for his encouragement, advice, and support in writing this book. I am also indebted to him for proofreading the manuscript. I also thank Professor Christian Pedersen for his critical reading of the final manuscript.

Mikael Bols

# INTRODUCTION

Stereoselectivity in organic synthesis is tremendously important because even moderately complex organic molecules cannot be prepared without reasonable control of stereochemistry. There is an increasing need for organic synthesis to be enantioselective as well. This is primarily due to the decreasing usefulness of racemic products in the pharmaceutical industry, caused by the fact that a racemic drug in a biological system behaves as a mixture of two different compounds of which only one has the desired properties, while the other is potentially harmful and, at least, undesirable.

Making a synthesis enantioselective can be a simple or difficult task, but regardless it requires introduction of optical activity from a natural source, either through a chiral auxiliary in an enantioselective reaction or by using a chiral starting material. Both strategies have advantages and limitations that make one or the other suitable in different cases.

The latter strategy is probably, at least for the time being, the simplest to plan or carry out if a useful chiral starting material can be found. But this is also its big limitation. Though thousands of optically active natural compounds are known, it is not always easy to select an available compound that would not require many steps to convert it into the desired structure.

Carbohydrates have been extensively used as starting materials in enantioselective synthesis because some are extremely inexpensive compounds. However, many of these syntheses start from glucose, and because glucose does not resemble the final target, these syntheses feature a tedious, multistep process of functional group conversion and protection/deprotection, making the strategy of employing carbohydrate starting materials not always very attractive.

Surprisingly, other inexpensive sugars and sugar derivatives are rarely used in synthesis, and, more importantly, only relatively few methods of converting the sugars to useful building blocks are employed. There are a number of readily available building blocks that can be prepared, in a few steps and without chromatographic separations, from carbohydrates that have never been used in synthesis. I believe this is because most synthetic chemists are not aware of these compounds, since they are difficult to locate in the literature. These compounds have usually been prepared by carbohydrate chemists who do not often venture outside their field and attempt to synthesize noncarbohydrate compounds.

The primary goal of this book is to make available a collection of building blocks from which the synthetic chemist can easily locate chiral starting materials that he or she would find useful. To make this more readily comprehensible, the synthesis and chemistry of the various types of building blocks will be covered in 10 short chapters. Which carbohydrate raw materials are inexpensively available will also be discussed, as well as how some expensive but very useful ones, such as 1,6-anhydro-glucose, can be prepared.

## WHY USE CARBOHYDRATES?

The simple fact that the least expensive chiral compound, sucrose, is a carbohydrate seems to suggest the use of carbohydrates as starting materials for synthesis of chiral compounds. Though sucrose itself has not yet proved very valuable in this respect, a number of other inexpensive sugars, first and foremost, glucose, recommend themselves. The price of these materials is so low that synthesis from them can be carried out on any scale without difficulty, and that low yield in an initial step is acceptable. The drawback with using carbohydrates is that they tend to be "overfunctionalized" for many purposes. However, this can also be an advantage that gives carbohydrates great potential and versatility. Where chiral centers in other areas of chemistry are usually preserved with great care, this need not be the case in carbohydrate chemistry. There too many chiral centers may well be the problem, and the best synthesis of a noncarbohydrate compound is, therefore, often a synthesis where unneeded chiral centers are eliminated quickly. Even for the synthesis of compounds with only one chiral center, use of a carbohydrate building block can often compete with other strategies. This was the case for the synthesis of L-carnitine, as shown in Schemes 1–3. The synthesis

**Scheme 1** L-Carnitine synthesis from L-arabinose

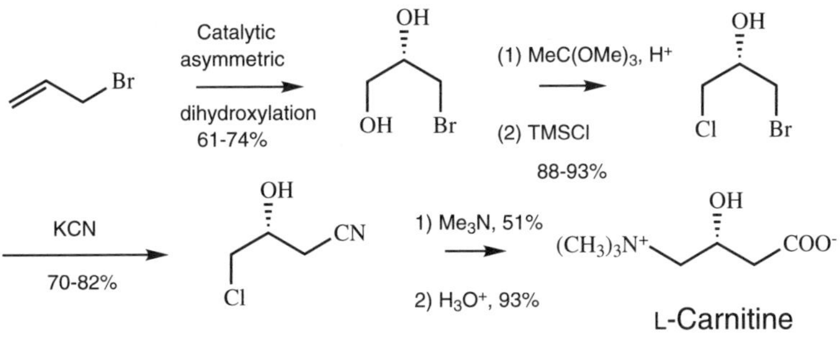

**Scheme 2** Catalytic asymmetric synthesis of L-carnitine

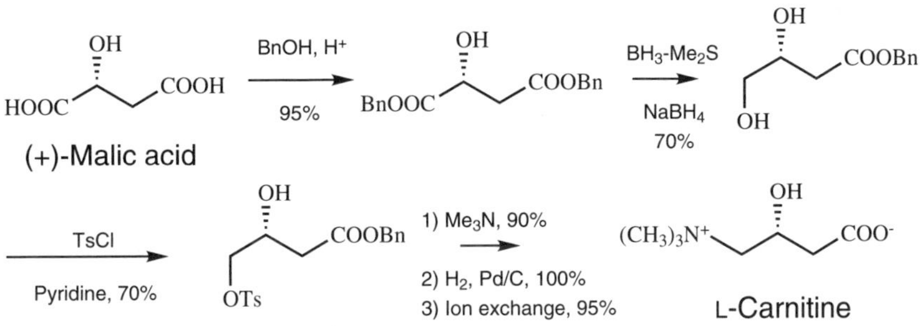

**Scheme 3** L-Carnitine synthesis from (+)-malic acid

of L-carnitine from L-arabinose[1] (Scheme 1) was neither longer nor less practical than either (a) a synthesis using enantioselective dihydroxylation[2] (Scheme 2) or (b) a synthesis from more "carnitine-looking," but expensive, (+)-malic acid[3] (Scheme 3).

## WHEN TO USE CARBOHYDRATES

It is certainly not the intention in this book to advocate uncritical use of carbohydrates as chiral building blocks, because there will obviously be many cases where their use is unwarranted. It can be very difficult, however, to determine when this is so. For example (Scheme 4), two often-used building blocks, **4.59** and its saturated analog, do not resemble carbohydrates. They can both be prepared from L-glutamic acid[4,5] or D-galactose.[6,7] Structure **4.59** is probably easiest to prepare from D-galactose, whereas the saturated lactone seems easiest to prepare from L-glutamic acid.

Generally speaking, the carbohydrates undoubtedly have their greatest usefulness as synthons for compounds containing carbon chains with contiguous or noncontiguous secondary alcohols. However, they have often been applied to the synthesis of any type of chiral compound. One feature that has often been taken advantage of in natural products synthesis from carbohydrates is the stereocontrol that can be obtained in the manipulation of functionalities in small rings. Be-

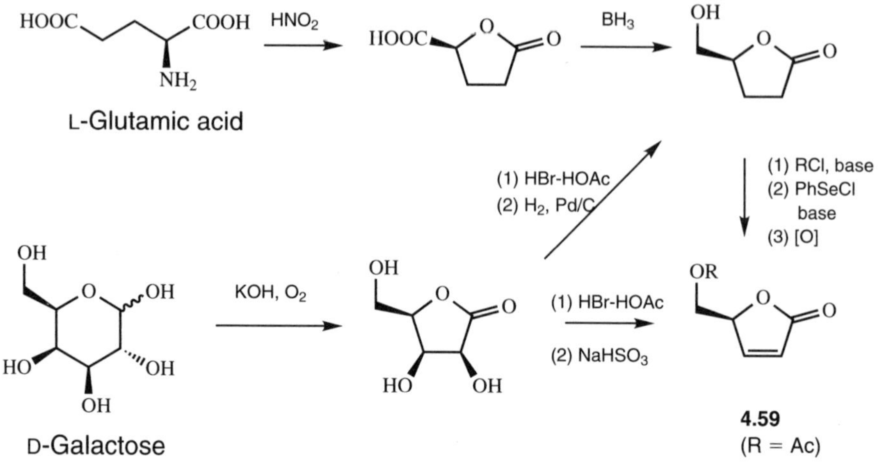

L-Glutamic acid

D-Galactose

4.59
(R = Ac)

Scheme 4

cause carbohydrate building blocks most often have a ring structure that easily can be opened at a later stage, a common strategy has been to carry out series of reactions on such building blocks to get, with good control of stereochemistry, compounds that only remotely resemble the parent sugar derivative. The stereocontrol obtained in transformations of small rings is typically very good. The preference for addition from the $\beta$ face in derivatives **2.07** and **2.06** has effectively been applied to the selective synthesis of branched compounds. For example, while hydroboration of **2.06** selectively gave compound **2.05** with the carbon substituent $\alpha$,[8] stereoselective addition of nitromethane to the ketone **2.07**[9] gave, after functional group interconversion,[10] compound **2.43** with the carbon substituent $\beta$ (Scheme 5). In this example the bulky 1,2-acetonide controls the

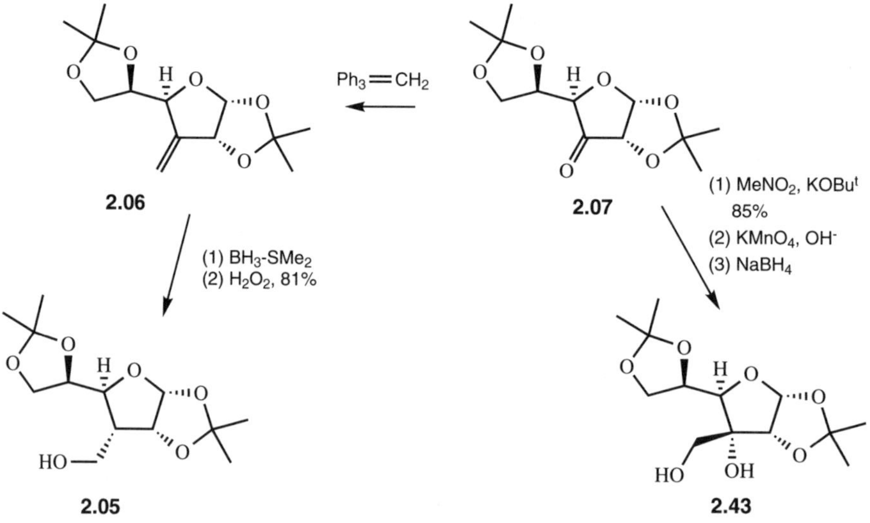

2.06

2.07

2.05

2.43

Scheme 5

4.57 $\xrightarrow{H_2,\ Pd/C}$

4.58 $\xrightarrow{H_2,\ Pd/C}$ 4.56

**Scheme 6**

stereochemistry by exerting steric hindrance from the $\alpha$ face. Much smaller substituents can, however, exert an equally profound influence (Scheme 6). Thus $\alpha,\beta$-unsaturated $\gamma$-lactones are hydrogenated with very high stereoselectivity to saturated lactones.[11] Even more surprisingly, the reaction works equally well on $\delta$-lactones.[12] In both cases the exocyclic carbon substituent exerts sterical hindrance on one side of the molecule so that addition of hydrogen takes place from the other side.

Examples of synthesis of compounds with remote carbohydrate resemblance from carbohydrates can be found in Hanessians book.[13] It is recommended that a retrosynthetic analysis be carried out so it can be determined, using the compendium of building blocks, if a desired synthon can be readily prepared from carbohydrates.

## HOW TO USE CARBOHYDRATES

To make effective use of carbohydrates in synthesis, it is important to benefit from the vast practical experience in the field. As an example, let us look at a very recent synthesis of a natural product[14] in which intermediate **A** was prepared (Scheme 7). Synthesis of **A** starts from D-arabinose and is quite logical. Because **A** contains three chiral centers but spreads out over a five-carbon chain, not all of them can conceivably come from a single naturally occurring carbohydrate (Scheme 8). Therefore only the two consecutive carbons are taken from a carbohydrate, and can be located as C-3 and C-4 in D-arabinose. Their isolation was then carried out by stepwise removal of the two superfluous hydroxy groups in 10–12 steps, followed by final chain extension and elaboration into **A**—for a total of 18 steps. Is this an optimum use of the potential of carbohydrates? Looking at the relative simple structure of **A**, one would instinctively say no. Yet how can we find a better building block than D-arabinose or its thioacetal relative? We will return to this problem later after an introduction to the use of this book.

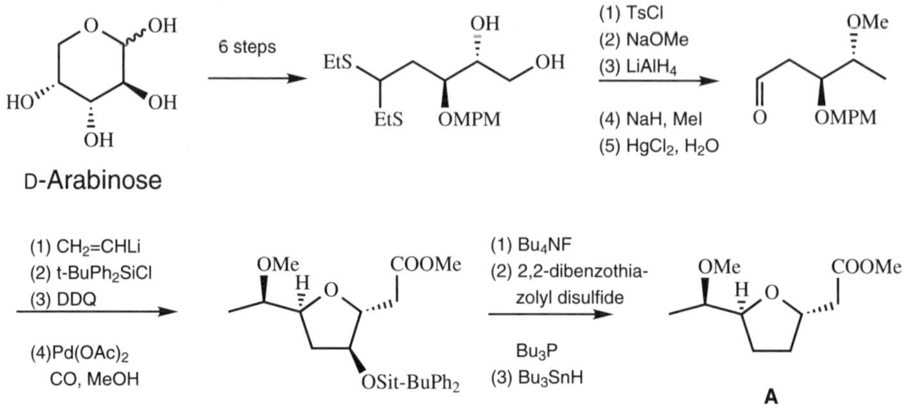

D-Arabinose

6 steps

(1) TsCl
(2) NaOMe
(3) $LiAlH_4$
(4) NaH, MeI
(5) $HgCl_2$, $H_2O$

(1) $CH_2$=CHLi
(2) t-$BuPh_2SiCl$
(3) DDQ
(4) $Pd(OAc)_2$
CO, MeOH

(1) $Bu_4NF$
(2) 2,2-dibenzothiazolyl disulfide
$Bu_3P$
(3) $Bu_3SnH$

**A**

**Scheme 7**

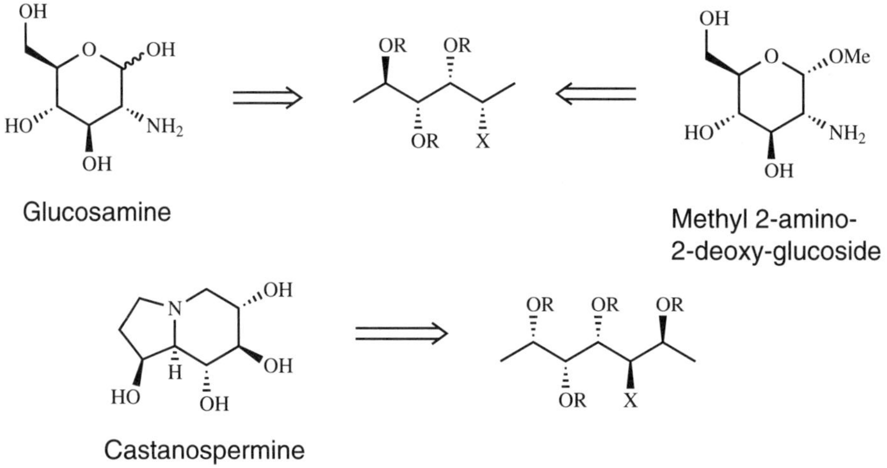

Glucosamine

Methyl 2-amino-2-deoxy-glucoside

Castanospermine

**Scheme 8**

## HOW TO USE THIS BOOK TO FIND BUILDING BLOCKS

This book contains compounds with up to eight consecutive chiral centers. All are listed in the stereochemical index according to chirality.

The first step is to identify chiral fragments in the target molecule that can conceivably come from one of these compounds—that is, fragments with a maximum of 10 carbons and eight consecutive chiral centers. To use the index the stereochemical sequence of such fragments is written in extended conformation, with the carbon chain as backbone and placing the substituent in three categories:

1. OR for hydroxyl groups, ethers, esters, epoxides, sulfonates
2. X for all heteroatoms other than oxygen (halides, amines)
3. C for carbon substituents (i.e., branched compounds)

The extended conformation structure is drawn taking the following rules into account with the priority listed:

1. Leftmost zigzag is pointing up
2. Achiral atoms as far as possible to the right
3. C substituents as far as possible to the right
4. X substituents as far as possible to the right
5. Leftmost substituent bond upwards (if possible)
6. Next leftmost substituent bond upwards (if possible)

Chiral centers that are acetals or otherwise labile are not taken into account. Some examples are given in Scheme 8.

Using these extended conformation formulae, each fragment is then looked for in the index. Compounds are listed according to increasing number of chiral centers. Under each number of chiral centers, compounds with consecutive chiral centers having exclusively oxygen functionalities are listed first. Then, compounds in which one or more oxygen substituents have been substituted, first with heteroatoms and then with carbon substituents, are listed. Finally, compounds with archiral atoms in the stereochemical sequence are listed. The carbohydrate nomenclature for each stereochemical sequence has been listed. It should be noted, however, that each sequence normally will have two names, depending on the reference structure. Thus the glucosamine-derived sequence in Scheme 8 can be termed either D-*gluco* or L-*gulo*.

To return to our example, let us try to find stereochemical fragments to compound **A** (Scheme 9). A stereochemical fragment containing all three chiral hydroxy groups can be written in extended conformation, with the leftmost apex pointing up, in two possible ways: **B** and **C**. However, **B** has the substituent-free atoms mostly to the right, so **B** is the one to be found in the index. **B** is not, however, found in the stereochemical index. If we search for only the two consecutive chiral centers (fragment **D**), a large number of building blocks are found. One of

OMe H O COOMe
**A**
OR OR OR
**B**
OR OR
**D**
OR OR OR
**C**
OR OR
OH H O O
**4.48**

**Scheme 9**

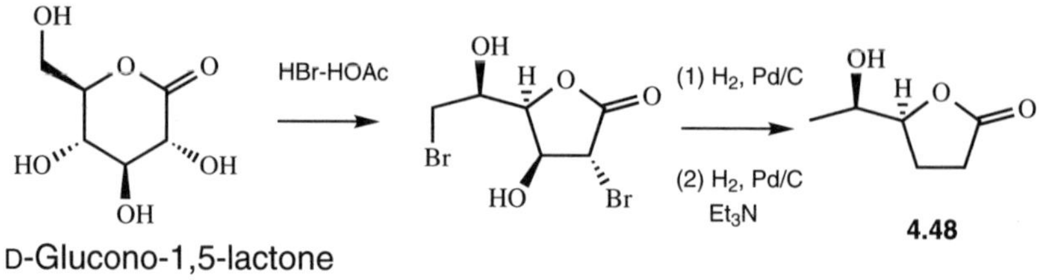
OH
O
O
HO
OH
OH
D-Glucono-1,5-lactone
HBr-HOAc
OH
H
O
O
Br
HO
Br
(1) H2, Pd/C
(2) H2, Pd/C
Et3N
OH
H
O
O
4.48

**Scheme 10**

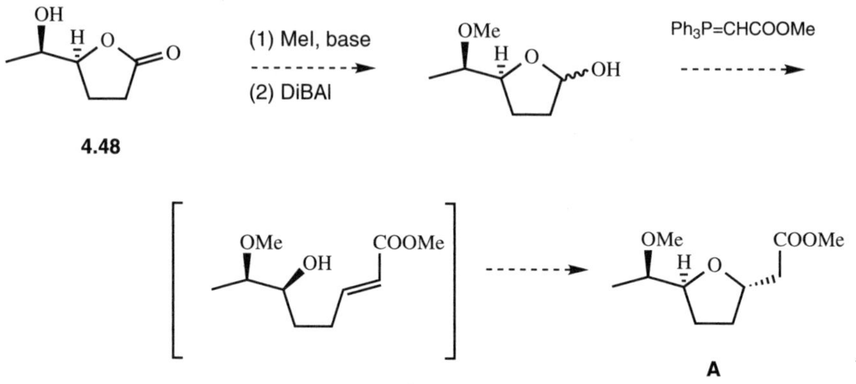
OH
H
O
O
4.48
(1) MeI, base
(2) DiBAl
OMe
H
O
OH
Ph3P=CHCOOMe
OMe
OH
COOMe
OMe
H
O
COOMe
A

**Scheme 11** Proposed synthesis of **A**

those is compound **4.48**. A look in the building-block compendium reveals that **4.48** can be prepared in three steps from D-glucono-1,5-lactone[6] (Scheme 10). Now methylation, reduction to the lactol, and a Wittig reaction with a stabilized ylid would lead to the (*E*)-2,3-unsaturated ester. Such 6-hydroxy-2,3-unsaturated esters are known[15] to undergo spontaneous cyclization under basic catalysis. This would then lead to **A** and its diastereomer (Scheme 11), with **A** probably predominant. The synthesis would have a total of six steps and would likely be a great improvement on the 18-step synthesis previously described. The last steps are hypothetical, but they are meant to illustrate how synthetic routes can be improved by finding the best starting building block.

## REFERENCES

1. K. Bock, I. Lundt, and C. Pedersen. *Acta Chem. Scand.* **1983** *B37*:341–344.
2. H.C. Kolb, Y.L. Bennani, and K.B. Sharpless. *Tetrahedron Asymmetry* **1993** *4*:133–141.
3. F. D. Bellamy, M. Bondoux, and P. Dodey. *Tetrahedron Lett.* **1990** *31*:7323–7326.
4. S. Hanessian and P. J. Murray. *Tetrahedron* **1987** *43*:5055–5072.
5. K. Tonioka, T. Ishiguro, Y. Iitake, and K. Koga. *Tetrahedron Lett.* **1980** *21*:2973–2976.

6. I. Lundt and C. Pedersen. *Synthesis* **1986** 1052–1054.
7. J. A. J. M. Vekemans, G. A. M. Franken, C.W. Dapperens, E. F. Godefroi, and G. J. F. Chittenden. *J. Org. Chem.* **1988** *53:*627–633.
8. A. Mazur, B. E. Tropp, and R. Engel. *Tetrahedron* **1984** *40:*3949–3956.
9. H. P. Albrecht and J. G. Mofatt. *Tetrahedron Lett.* **1970** 1063–1066.
10. W. P. Blackstock, C. C. Kuenzle, and C. H. Eugster. *Helv. Chem. Acta* **1974** *57:*1003–1009.
11. L. F. Sala, A. F. Cirelli, and R. M. de Lederkremer. *Carbohydr. Res.* **1980** *78:*61–66.
12. O. J. Varela, A. F. Cirelli, and R. M. de Lederkremer. *Carbohydr. Res.* **1979** *70:*27–35.
13. S. Hanessian. *Total Synthesis of Natural Products. The Chiron Approach.* Pergamon Press, Oxford, **1983**.
14. M. F. Semmelhack, W. R. Epa, A.W.-H. Cheung, Y. Gu, C. Kim, N. Zhang, and W. Lew. *J. Am. Chem. Soc.* **1994** *116:*7455–7456.
15. D. Horton and D. Koh. *Carbohydr. Res.* **1993** *250:*231–247.

# 1

# THE RAW MATERIALS

As this book will show, despite the large number of useful carbohydrate compounds and derivatives known, there are, somewhat surprisingly, only about two dozen carbohydrate or carbohydrate-derived compounds that can be termed very inexpensive. The explanation for this is that only one carbohydrate-based synthesis, the synthesis of ascorbic acid, is performed on an industrial scale and thus provides inexpensive synthetic intermediates. Most of the other inexpensive carbohydrates are obtained directly from natural sources. An amusing consequence of this phenomenon is that diacetone ketogulonic acid (**1.22**), an intermediate in ascorbic acid synthesis, costs less than half the price of diacetoneglucose (**2.01**), even though its synthesis from glucose is much more lengthy than that of the latter substance. (Scheme 1.1) All building blocks described in this book are derived from these two dozen inexpensive compounds. Generally only compounds that can be prepared in four or fewer steps will be mentioned unless they are considered of exceptional synthetic value.

A fair number of carbohydrate compounds and building blocks other than those already mentioned are available at a higher price range, and the reader is

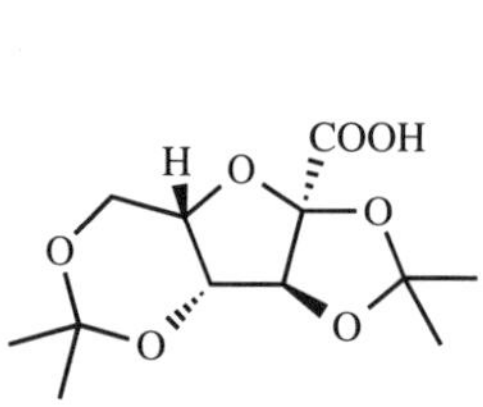

**1.22** 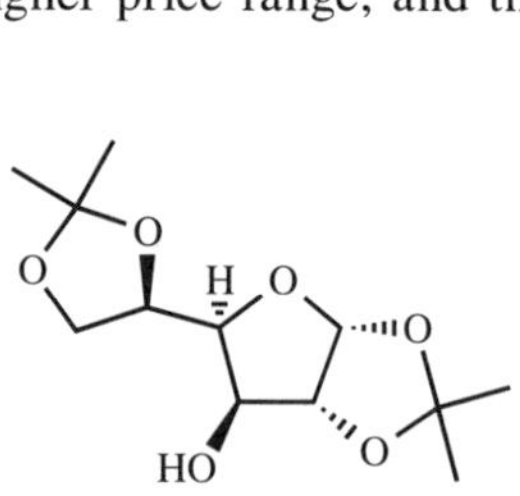 **2.01**

**Scheme 1.1**

encouraged to take advantage of this if he or she can justify it. Synthesis of these compounds are also described here.

## THE INEXPENSIVE CARBOHYDRATES

The approximate prices (1995) of the least expensive carbohydrate derivatives at the fine chemical suppliers are listed in Table 1.1. Though such prices little reflect the actual or potential manufacturing cost of these compounds, they are important to the chemist who wishes to use a carbohydrate as starting material for his or her synthesis. A list of industrial prices of carbohydrates available in bulk is found in Table 1.2.

The general natural sources of carbohydrates are various polysaccharides such as starch, mannans, and xylans. An exception is the largest carbohydrate product, sucrose (**1.23**), which is obtained directly from sugar cane or sugar beet and is therefore very inexpensive. Because of its complicated structure, however, su-

**Table 1.1 Approximate prices of carbohydrate derivatives (1995)**

| Compound | Price ($/100 g) | Source |
|---|---|---|
| Sucrose (**1.23**) | 0.5 | Sugar cane |
| D-Glucose (**1.01**) | 0.5 | Starch |
| D-Fructose (**1.09**) | 1.0 | Glucose |
| D-Gluconic acid (**1.12**) | 1.0 | Glucose |
| D-Glucitol (sorbitol) (**1.19**) | 1.0 | Glucose |
| Lactose (**1.24**) | 1.5 | Milk (whey) |
| D-Mannitol (**1.20**) | 2.0 | Fructose |
| Methyl $\alpha$-glucopyranoside (**1.11**) | 4.0 | Glucose |
| Maltose (**1.25**) | 4.0 | Starch |
| D-Isoascorbic acid (**1.16**) | 4.0 | Glucose |
| D-Glucono-1,5-lactone (**1.12**) | 5.0 | Glucose |
| D-Galactose (**1.02**) | 6.0 | Lactose |
| L-Sorbose (**1.10**) | 7.0 | Glucose |
| D-*Glycero*-D-*gulo*-heptonic acid (**1.13**) | 7.0 | Glucose |
| D-Glucosamine (**1.18**) | 10 | Sea shells |
| D-Xylose (**1.07**) | 10 | Wood |
| Dianhydroglucitol (**1.21**) | 12 | Glucose |
| D-Glucurono-3,6-lactone (**1.14**) | 13 | Glucose |
| L-Ascorbic acid (**1.15**) | 14 | Glucose |
| L-Arabinose (**1.05**) | 33 | Plant gum |
| D-Arabinose (**1.04**) | 33 | Glucose |
| Diisopropylideneketogulonic acid (**1.22**) | 40 | Glucose |
| D-Ribose (**1.06**) | 44 | Yeast |
| D-Mannose (**1.03**) | 46 | Ivory nut |
| D-Glucaric acid (**1.17**) | 57 | Glucose |
| L-Rhamnose (**1.08**) | 125 | Oak bark |

**Table 1.2 Approximate prices of carbohydrates in bulk**

| Compound | Price ($/kg) |
|---|---|
| Lactose (**1.24**) | 0.5–1.5 |
| Sucrose (**1.23**) | 0.8 |
| D-Glucose (**1.01**) | 0.6–1.1 |
| D-Fructose (**1.09**) | 0.9 |
| Sodium D-gluconate (**1.12**) | 1.5 |
| D-Glucitol (sorbitol) (**1.19**) | 1.6 |
| D-Mannitol (**1.20**) | 7.3 |
| L-Ascorbic acid (**1.15**) | 17 |

*Source: Chemical Marketing Reporter.* Nov. 4, 1994.

crose is not very useful as a building block or for the production of other sugars. Enzymatic hydrolysis with invertase to glucose (**1.01**) and fructose (**1.09**), and enrichment of the fructose content with glucose isomerase, which isomerize glucose (**1.01**) to fructose (**1.09**), allows production of fructose (Scheme 1.2). Pure D-fructose (**1.09**) is isolated by chromatography, or the mixture can be reduced with $H_2$ to give D-mannitol (**1.19**) and D-glucitol (**1.20**). The former can be isolated by crystallization.

**Scheme 1.2** Industrial production of D-mannitol

D-Glucose (**1.01**) and D-glucitol (**1.19**) are not prepared in this way. Glucose (**1.01**) is obtained from enzymatic or acidic hydrolysis of starch, and hence gluci-

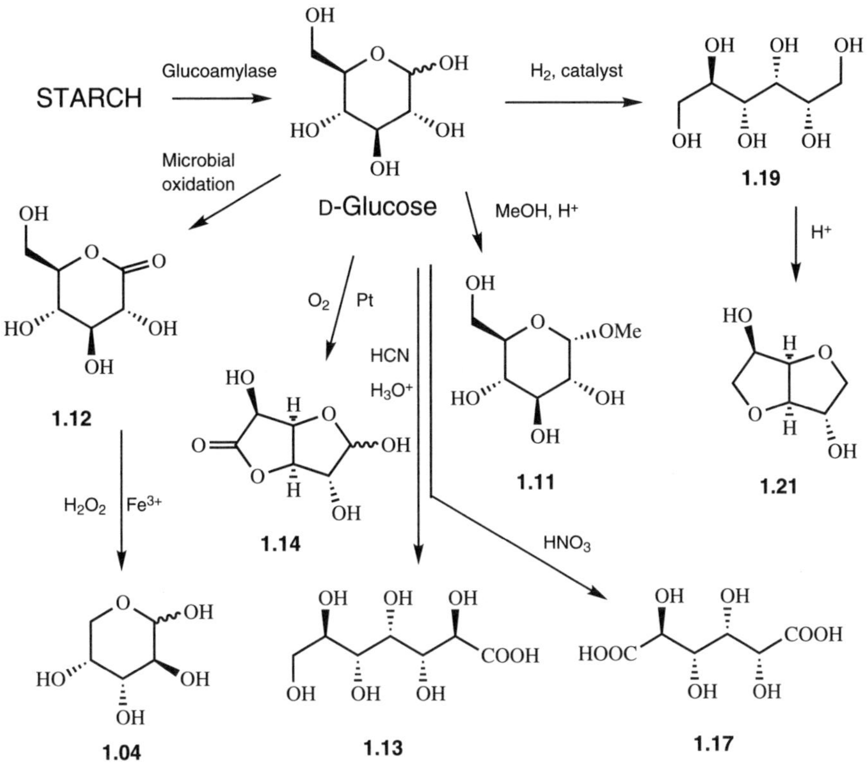

**Scheme 1.3** Commercial products from D-glucose

tol is obtained by hydrogenation of glucose. A number of other inexpensive compounds are obtained by one-step synthesis from these two compounds (Scheme 1.3). All these compounds are useful but it has to be kept in mind that they, with the exception of **1.04** and **1.13** have the same stereochemistry as glucose.

More interesting are the compounds obtained from the industrial synthesis of L-ascorbic acid (Scheme 1.4). They are L-sugars and can often be converted to fragments that are enantiomers of those obtained from other carbohydrates. L-Ascorbic acid is especially valuable.

Seven other hexoses and pentoses are readily avaiable from natural sources and are important alternatives to glucose and glucose-derived compounds.

D-Galactose (**1.02**) is obtained by hydrolysis of lactose (**1.24**) and removal of glucose by chromatography or fermentation. Lactose (**1.24**) is readily obtained from the waste product whey (Scheme 1.5).

D-Mannose (Scheme 1.6) is obtained by acidic hydrolysis of polysaccharides, mannans, present in ivory nut[1] or African doum palm kernel.[2] Similarly, L-arabinose (**1.05**) and D-xylose (**1.07**) are obtained by acidic hydrolysis of polysaccharides in plant gums and wood, respectively. L-Rhamnose (**1.08**) is obtained

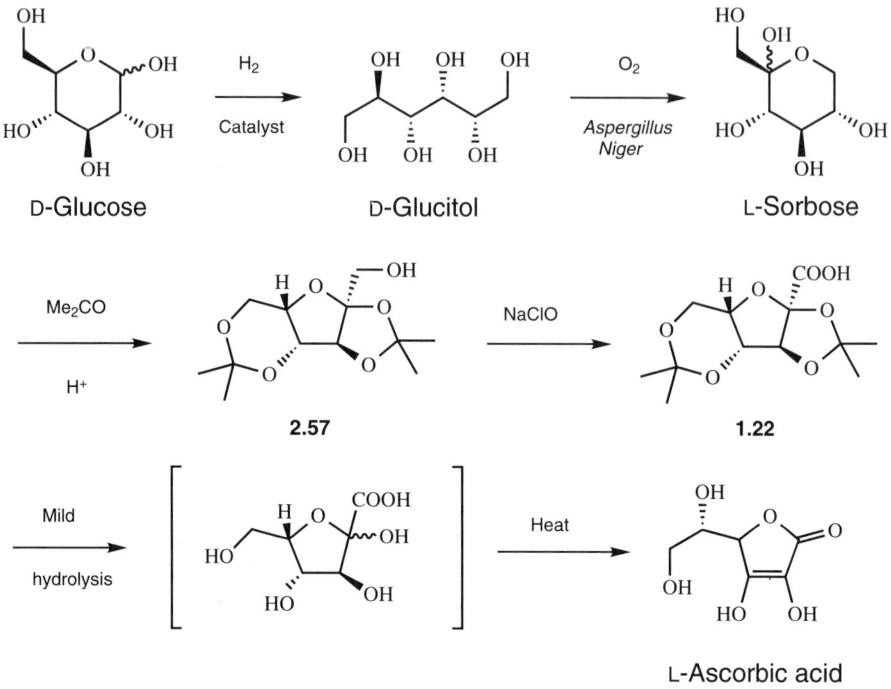

D-Glucose → ($H_2$, Catalyst) → D-Glucitol → ($O_2$, *Aspergillus Niger*) → L-Sorbose

→ ($Me_2CO$, $H^+$) → **2.57** → (NaClO) → **1.22**

→ (Mild hydrolysis) → [ ] → (Heat) → L-Ascorbic acid

**Scheme 1.4** Industrial synthesis of L-ascorbic acid

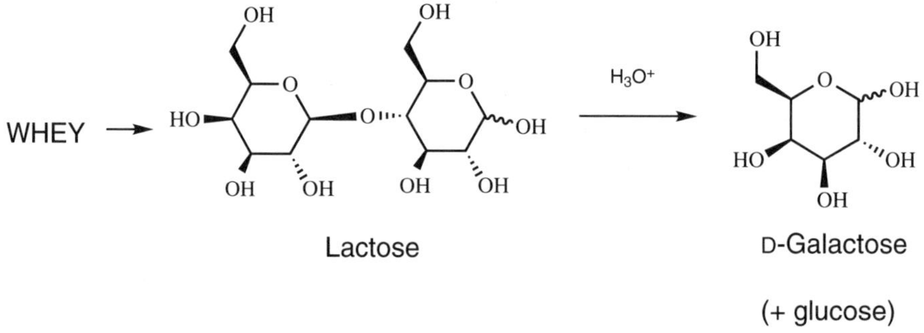

WHEY → Lactose → ($H_3O^+$) → D-Galactose (+ glucose)

**Scheme 1.5** Production of D-galactose

from the bark of an oak species. D-Ribose (**1.06**) is probably most often made by acid hydrolysis of the nucleoside fraction from yeast, though a chemical synthesis involving oxidation of D-arabinose to the acid, followed by base-catalyzed epimerization to D-ribonic acid and reduction, is claimed to be cheaper. The nitrogen-containing monosaccharide glucosamine (2-amino-2-deoxy-D-glucose, **1.18**, Scheme 1.6) is also inexpensive. It is obtained by acid hydrolysis of chitin, a polysaccharide present in sea shells.[3]

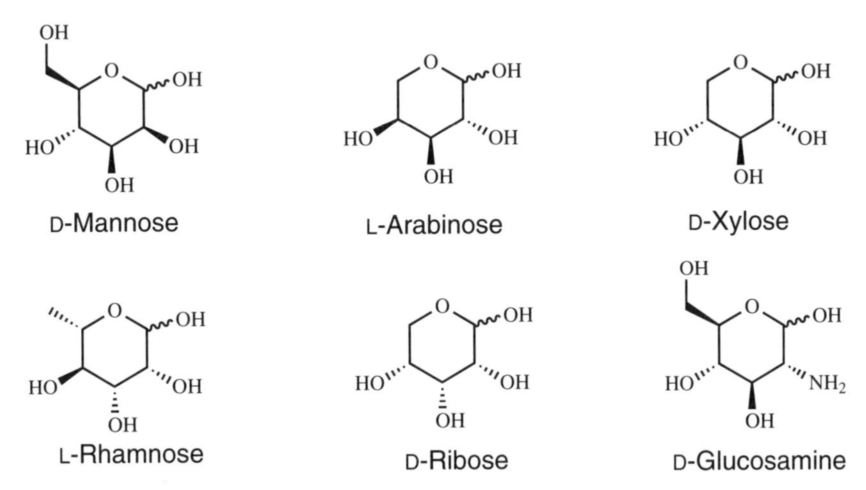


**Scheme 1.6** Sugars obtained from plants or animals

## THE ROUTES TO THE BUILDING BLOCKS

For synthetic purposes, many other, especially simpler stereochemical sequences are needed than present in these basic raw materials. A large number of reaction pathways have therefore been employed to convert these raw materials into compounds with different stereochemical sequences or with a specific selective protection of the functionalities. These pathways show a great diversity and will be dealt with in detail in the following 10 chapters. However, some common trends are found. Scheme 1.7 shows 10 of the most common stereochemical sequences found in the compounds compiled here and also shows where they come from. All the sequences shown have consecutive chiral centers with oxygen functionalities. For the sequences with two or three chiral centers, all possible stereoisomers are shown while only the three most abundant with four chiral centers are shown. The number below each sequence is its relative abundance among compounds found. The arrows and the percentages show from what source the indicated fraction of compounds with the sequence was obtained. For example, 56% of the compounds with the leftmost sequence are obtained from L-arabinose. The sources of the other 44% of the compounds are not shown, because each typically contribute less than 20%. Scheme 1.7 reveals clearly the large extent D-glucose is being used as raw material. This is partly because of the low price of D-glucose, but also because of its extensively developed chemistry. D-Glucose is the largest contributor to all the six sequences that can possibly be obtained from it without configurational inversions. Even compounds with D-xylose stereochemistry are more often obtained from D-glucose than from inexpensive D-xylose itself. However, Scheme 1.7 also reveals that for synthesis of compounds with sequences that would require configurational inversions to be made from D-glucose, glucose is almost never used and such compounds are less abundant. Compounds con-

**Natural Sources**

| L-Arabinose | D-Galactose | D-Xylose | **D-Glucose** | D-Fructose | D-Mannose | D-Ribose |
|---|---|---|---|---|---|---|
| **1.05** | **1.02** | **1.07** | **1.01** (incl. **1.04, 1.11-1.14, 1.17, 1.19, 1.21**) | **1.09** (incl. **1.20**) | **1.03** | **1.06** |

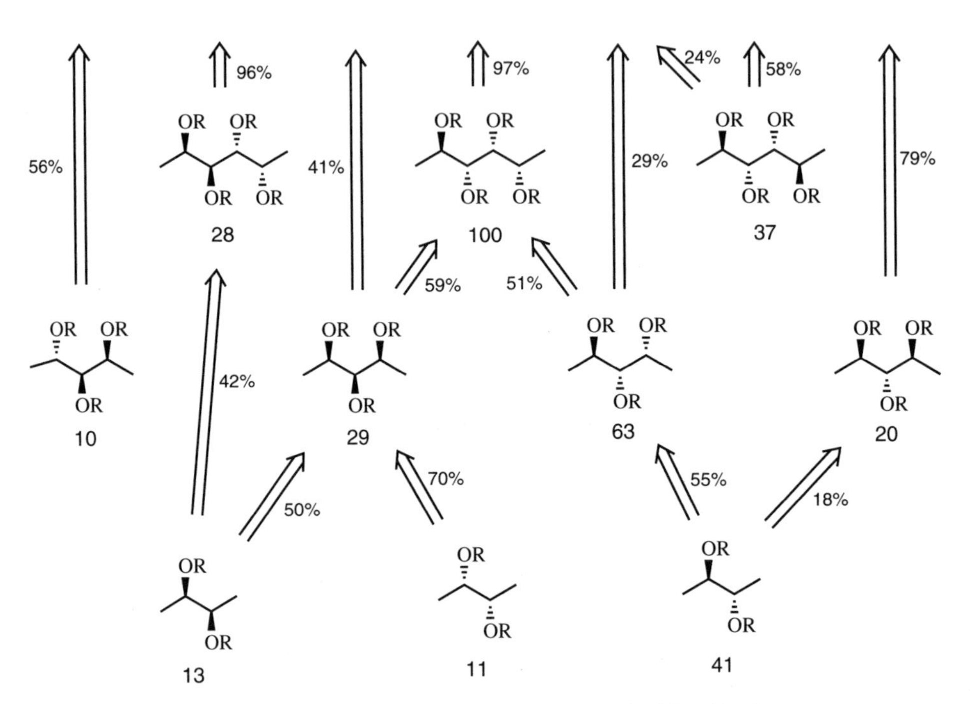


**Scheme 1.7** Pathways to the most common building blocks

taining D-ribose stereochemistry are almost exclusively obtained from D-ribose. The same is true to a lesser extent for L-arabinose. This fact emphasizes the unique importance of those sugars. It is also interesting to note that while compounds with D-galactose or D-glucose stereochemistry are exclusively obtained from their parent sugar, mannose stereochemistry is to a large extent supplied from fructose in the form of D-mannitol.

In Table 1.3 we see the relative abundance of all the building blocks with four consecutive chiral centers. It should be noted that because compounds have been collected with the emphasis on short reaction routes and inexpensive precursors, it cannot be concluded that certain chiral sequences cannot be obtained from carbohydrates. Thus, the fact that there are no compounds with the L-altro or L-talo sequence does not mean that L-altrose or L-talose has not been made from carbohydrates. They have. But it does indicate how easily the various stereochemistries can be obtained.

**Table 1.3 Relative abundance of building blocks with four chiral atoms**

| Sequence | Name | C–O | C–X[a] | C–C[a] |
|---|---|---|---|---|
| 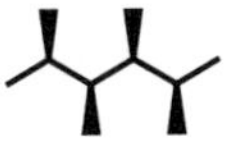 | D-Ido- | 3 | 1 | 0 |
| 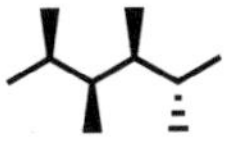 | D-Gulo-/L-gluco- | 3 | 1 | 0 |
| 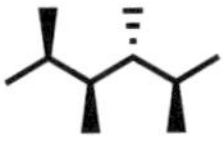 | D-Altro-/D-talo | 8 | 3 | 2 |
| 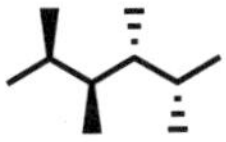 | D-/L-Galacto | 28 | 4 | 0 |
| 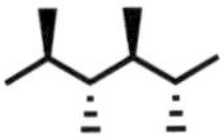 | D-/L-Allo | 5 | 3 | 5 |
| 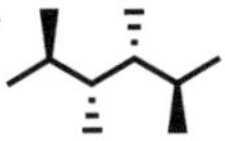 | D-Manno- | 37 | 5 | 0 |
| 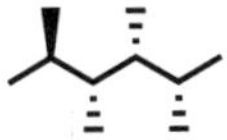 | D-Gluco-/L-gulo- | 100 | 6 | 5 |
| 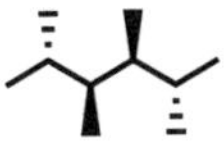 | L-Manno- | 9 | 0 | 0 |
| 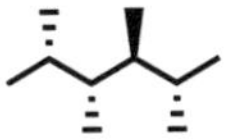 | L-Altro-/L-talo- | 0 | 1 | 0 |
| 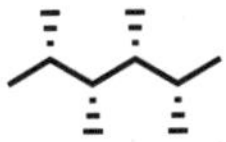 | L-Ido- | 3 | 1 | 0 |

[a]One or more C–O bonds exchanged with this bond.

Compounds with interruptions in the sequence of chirality (i.e., which have achiral atoms in the sequence) show a similar trend in that they have been primarily made from D-glucose. The sequences obtained by elimination of the chirality of C-3 or C-4 of a hexose are shown in Scheme 1.8. Six of the possible eight sequences are readily available, and four of them have been made from D-glucose. They have all been made by elimination of chirality at C-3 or C-4 in glucose, mannose, galactose, or rhamnose except for one, which was made by double

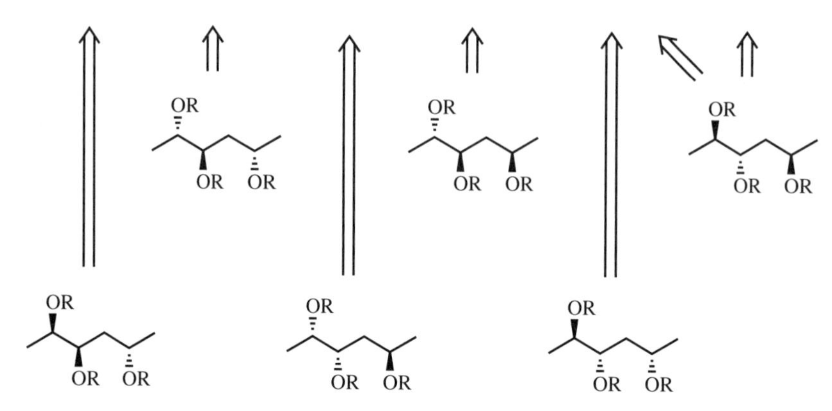


**Scheme 1.8** Pathways to interrupted sequences

inversion of two chiral centers in glucose (compound **4.71**). Again, configurational inversion seems to be unusual in the effective synthesis of these sequences.

## REFERENCES

1. H. S. Isbell and H. L. Frush. *Methods Carbohydr. Chem.* **1962** *1*:145–147.
2. M. A. E. Sallam. *Acta Chem. Scand.* **1977** *B31*:181–182.
3. M. Stacey and J. M. Webber. *Methods Carbohydr. Chem.* **1962** *1*:228–230.

# 2

# CARBOHYDRATE ACETAL DERIVATIVES

Because a monosaccharide such as glucose contains both an aldehyde and many hydroxy groups, the possibilities of forming acetal/ketal derivatives are many. This is traditionally the most extensively employed method of selectively manipulating the functionalities of the carbohydrate.

## METHYL GLYCOSIDES

Though an aldehyde, the monosaccharide is practically exclusively in cyclic hemiacetal form. Reaction with methanol (or another alcohol) and acid converts it into a monomethyl acetal, a methyl glycoside (Scheme 2.1). This reaction, the so-called Fisher glycosidation,[1] is a complicated reaction that can lead to four isomeric products. However, in many cases, the most thermodynamically stable product, typically the $\alpha$-pyranoside, can be isolated in good yield by simple crystallization. Sometimes a short reaction time allows isolation of a five-membered kinetic product, a furanoside (Scheme 2.2).[2,3]

These methyl glycosides are useful because the cyclic form of the sugar is fixed and they are stable to many reagents. They do require prolonged acid hydrolysis to be reconverted to monosaccharide[4] (Scheme 2.3).[5]

## CYCLIC ACETALS

Given the many hydroxy groups in carbohydrates, their reaction with aldehydes or ketones under acid catalysis seems *a priori* to be no simple matter. The following guidelines can be given:

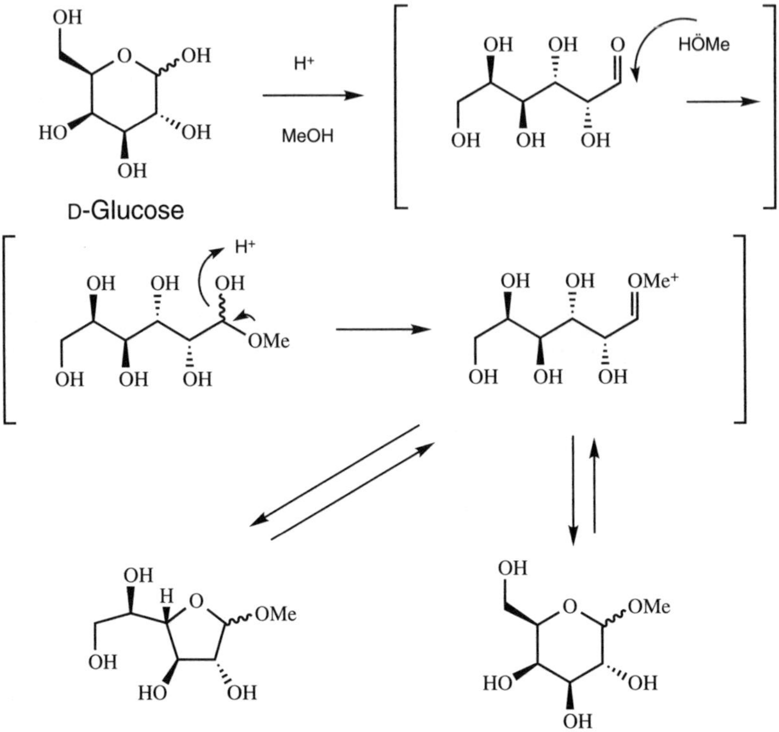

$H^+$

MeOH

HÖMe

D-Glucose

$H^+$

OMe

$OMe^+$

Furanosides
(kinetic products)

Pyranosides
(thermodynamic products)

**Scheme 2.1**

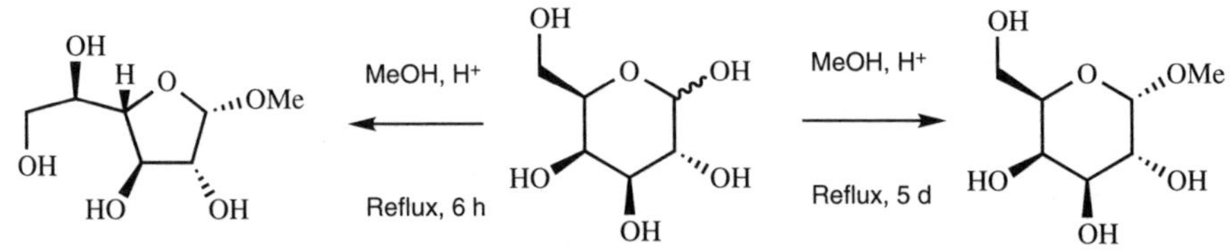

MeOH, $H^+$

Reflux, 6 h

MeOH, $H^+$

Reflux, 5 d

68% by chromatography

D-Galactose

37% by crystallization

**Scheme 2.2**

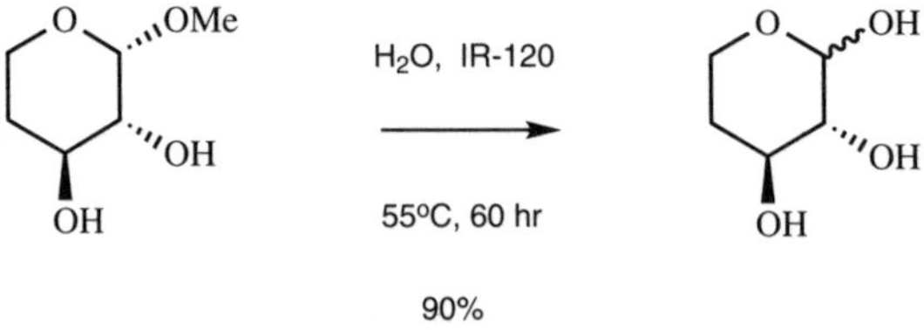

$H_2O$, IR-120

55°C, 60 hr

90%

**Scheme 2.3**

1. Aldehydes preferably form six-membered cyclic acetals.
2. Ketones preferably form five-membered cyclic ketals.
3. A five-membered cyclic ketal fused to a five-membered ring is particularly stable and favorable.

Some examples are shown in Scheme 2.4.[6–9] Note that while monosaccharides can change cyclic structure during the reaction, methyl glycosides are fixed under acetalization conditions. Also note that an exocyclic acetonide can be hydrolyzed in the presence of an endocyclic fused to a furanose ring. A more comprehensive coverage of cyclic acetal formation on carbohydrates can be found elsewhere.[10–13]

D-Glucose → ($Me_2CO$, $H^+$, 91%) → **2.01** → ($H_2O$, $H^+$, 95%) → **2.03**

D-Glucose → ($PhCHO$, $ZnCl_2$, 42%) → **2.51**

L-Rhamnose → ($Me_2CO$, $H^+$, 68%) → **2.24**

Methyl L-rhamnoside → ($Me_2CO$, $H^+$, 95%) → **2.17**

**Scheme 2.4**

There are almost endless possibilities of manipulations of the free hydroxy groups of such partially protected derivates to make synthons. A number of these have been included in the building block compendium. Worth emphasizing is the use of oxidizing a hydroxy group to the ketone. Addition to the ketone is then often stereoselective. Thus, reduction can lead to configurational inversion (Scheme 2.5).[14–15]



**Scheme 2.5**

## REACTIONS OF CYCLIC ACETALS

Besides being useful for selective protection of monosaccharides, a number of interesting reactions can be carried out with cyclic acetals. This is particularly true for 4,6-benzylidene acetals, which can be selectively opened with *N*-bromosuccinimide to give a bromobenzoate (**2.56**)[16] and can be reductively opened with various reagents to give either a primary (**2.61**)[17] or a secondary benzyl group (**2.60**)[18] (Scheme 2.6), depending on the reducing agent used. $LiAlH_4/AlCl_3$ gives **2.60**, while $NaB(CN)H_3$ gives **2.61**.

Another useful reaction is Klemer's base fragmentation of a 2,3-benzylidene acetal to give ketone **2.55**[19] (Scheme 2.7).

Acetonides in some cases can be caused to migrate, as shown in Scheme 2.8.[20]

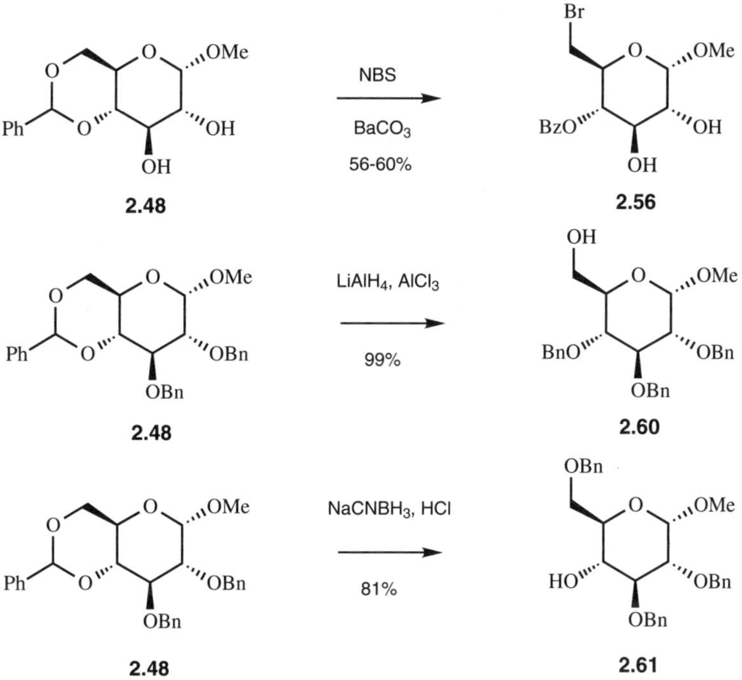

NBS, $BaCO_3$, 56-60%: **2.48** → **2.56**

$LiAlH_4$, $AlCl_3$, 99%: **2.48** → **2.60**

$NaCNBH_3$, HCl, 81%: **2.48** → **2.61**

**Scheme 2.6**

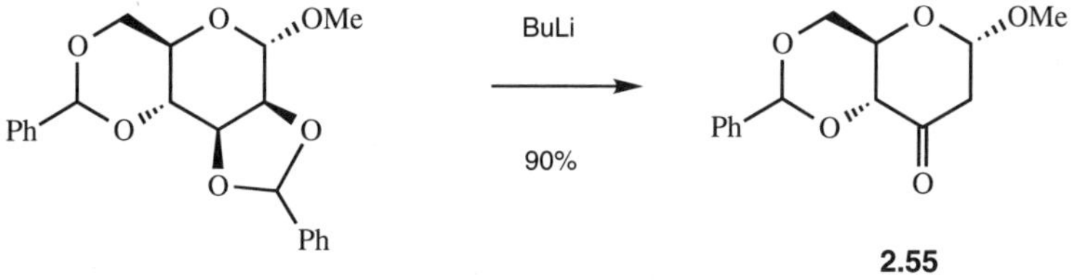

BuLi, 90% → **2.55**

**Scheme 2.7**

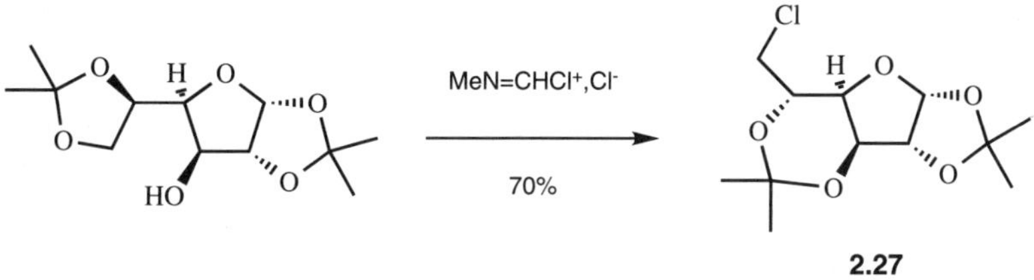

$MeN{=}CHCl^+,Cl^-$, 70% → **2.27**

**Scheme 2.8**

## CLEAVAGE OF CYCLIC ACETALS

Cyclic acetals of carbohydrates can be cleaved in the usual ways, typically acidic hydrolysis, which can be carried out without affecting a methyl glycoside. An interesting mild alternative is transacetalization with $I_2$/MeOH.[21]

## REFERENCES


1. A. F. Bochkov and G. E. Zaikov. *Chemistry of the O-Glycosidic Bond.* Pergamon Press, Oxford, **1979**, pp. 11–16.
2. I. Augestad and E. Berner. *Acta Chem. Scand.* **1954** *8:*251–256.
3. T. Skovby, L. Nørgård, and M. Bols (unpublished). D. F. Mowery, Jr. *Methods Carbohydr. Chem.* **1963** *2:*328–331.
4. A. F. Bochkov and G. E. Zaikov. *Chemistry of the O-Glycosidic Bond.* Pergamon Press, Oxford, **1979**, pp. 177–201.
5. K. E. Dow, R. S. Riopelle, W. A. Szarek, M. Bols, E. R. Ison, J. Plenkiewicz, A. Lyon, and R. Kisilevsky. *Biochem. Biophys. Acta* **1992** *1156:*7–14.
6. O. T. Schmidt. *Methods Carbohydr. Chem.* **1963** *2:*318–325.
7. H. G. Fletcher. *Methods Carbohydr. Chem.* **1963** *2:*307–308.
8. B. R. Baker. *Methods Carbohydr. Chem.* **1963** *2:*441–444.
9. J. Jary, K. Capek, and J. Kovar. *Collect. Czech. Chem. Commun.* **1963** *28:*2171–2181.
10. A. N. de Belder. *Adv. Carbohydr. Chem. Biochem.* **1965** *20:*220–302.
11. A. N. de Belder. *Adv. Carbohydr. Chem. Biochem.* **1977** *34:*179–242.
12. J. Gelas. *Adv. Carbohydr. Chem. Biochem.* **1981** *39:*71–156.
13. R. F. Brady, Jr. *Adv. Carbohydr. Chem. Biochem.* **1971** *26:*197–278.
14. J. D. Stevens. *Methods Carbohydr. Chem.* **1972** *6:*123–128.
15. S. M. Mio, Y. Kumagawa, and S. Sugai. *Tetrahedron* **1991** *47:*2133–2144.
16. S. Hanessian. *Methods Carbohydr. Chem.* **1972** *6:*183–189.
17. P. J. Garegg and H. Hultberg. *Carbohydr. Res.* **1981** *93:*C10–C12.
18. K. Umemura, H. Matsuyama, M. Kobayshi, and N. Kamigata. *Bull. Chem. Soc. Japan* **1989** *62:*3026–3028.
19. A. Klemer and G. Rodemeyer. *Chem. Ber.* **1974** *107:*2612–2614.
20. S. Hanessian. *Methods Carbohydr. Chem.* **1972** *6:*190–193.
21. W. A. Szarek, A. Zamojski, K. N. Tiwari, and E. R. Ison. *Tetrahedron Lett.* **1986** *27:*3827–2830.

# 3

# OTHER SELECTIVELY PROTECTED SUGARS

Besides use of acetals there are a number of other possibilities of selective functional group conversion of the different hydroxyl groups in a monosaccharide. These methods take advantage of the different chemical properties of the OH groups. A few of these properties are outlined in Scheme 3.1. There are basically two main ways of selective manipulation of the monosaccharide: selective protection or substitution. In the former the different nucleophilicity of the hydroxyl groups is used. In the latter all the hydroxyl groups are normally, though not always, converted to some kind of leaving group, and then the difference in leaving group ability is exploited. Both methods can be employed for making many useful building blocks.

least hindered OH, can be protected selectively

most reactive OH to substitution

least reactive OH to substitution

least hindered OH, can be protected selectively

R = H or alkyl

**Scheme 3.1**

## SELECTIVE PROTECTION

Because selective protection almost always consists of a nucleophilic attack of a monosaccharide OH group on an electrophilic reagent, two factors control the selectivity of such a process: the steric hindrance around the various hydroxyl groups and their nucleophilicity. Monosaccharides in both pyranoside and furanoside form can be selectively protected at the primary position in good yield with many different protection groups (Scheme 3.2).[1] This is to be expected because monosaccharide OH groups, especially the secondary ones, are less nucleophilic than alcohols in general, and definitely more sterically hindered that the primary alcohol. Almost always the primary hydroxyl group will be the first to be protected; however, there are methods that allow selective protection of the 1-OH.

TsCl

Pyridine
2 days, 20°C
55%

3.03

**Scheme 3.2**

It is possible to selectively esterify the anomeric position of glucose using acyltriazoles in aqueous solution, although yields are not high and separation from unreacted glucose is troublesome (Scheme 3.3).[2]

D-Glucose

Benzoyltriazole

Pyridine, NaH (cat.), 4 hr, 0°C
60% (based on benzoyltriazole)

**Scheme 3.3**

Making rules for the selective protection of the other secondary hydroxyl groups is more difficult because their reactivities depend on the configuration.[3] The difference in reactivity is also often very small, and other factors can also play a role. The 2-OH is generally the more reactive in methyl-$\alpha$-D-glycopyranosides, at least toward esterification whereas the 4-OH is the least reactive (Schemes 3.4 and 3.5).[4–6] However toward glycosidation of glucosides[7] the reacivity order is: 6-OH >> 3-OH > 2-OH > 4-OH. This again is in contrast to benzylation of methyl-$\alpha$-D-glucopyranoside, which occurs with surprising selectivity[8] to give the 2,4,6-tri-*O*-benzyl ether **3.64** (Scheme 3.6). These differences are

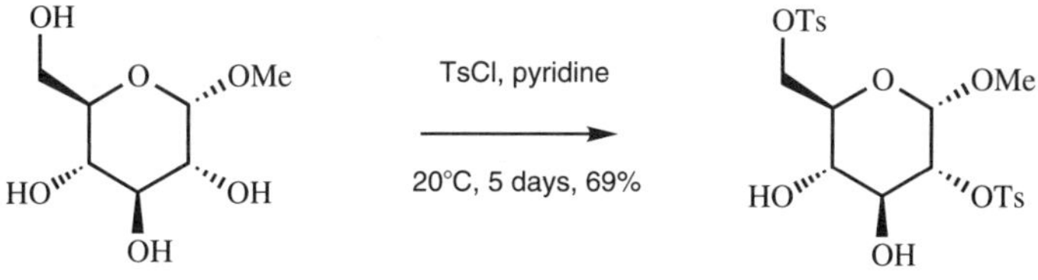

TsCl, pyridine

20°C, 5 days, 69%

**3.07**

**Scheme 3.4**

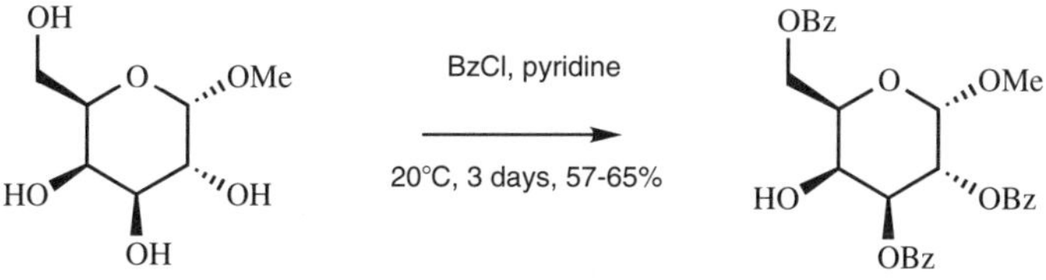

BzCl, pyridine

20°C, 3 days, 57-65%

**3.08**

BzCl, pyridine

20°C, 2 days, 57%

**3.09**

**Scheme 3.5**

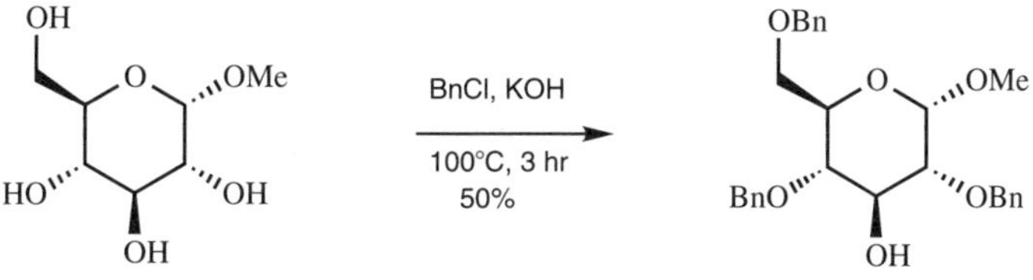

BnCl, KOH

100°C, 3 hr 50%

**3.64**

**Scheme 3.6**

undoubtedly due to differences in reaction conditions. The glycosidation reactions are alkylations occurring at acidic or neutral conditions, which means that the nucleophilicity of the alcohol is the important factor. In the benzylation reaction in which strong base is used, the nucleophilicity of the alcoxide is the important factor.

*In situ* preparation of tin ethers or tin acetals of the OH groups and reaction of these often increase yields and selectivity considerably.[9] With tin acetals, selectivity is governed by where the tin acetal can be formed. Thus, the dibutyltin acetal of methyl-α-D-galactopyranoside can be selectively monobenzylated in the 3-position to give **3.71** (Scheme 3.7).[10] This is explained by the preferential formation of a cyclic tinacetal between the 3- and 4-OH.

(1) $Bu_2SnO$ PhH, rfx
(2) BnBr, $Bu_4NI$ rfx, 3 hr
71%

**3.71**

**Scheme 3.7**

Esterification of reducing sugars is complicated by the fact that mutarotation can take place so that different ring forms can be obtained. A particularly interesting example on this phenomenon is benzoylation of fructose (Scheme 3.8), which can lead to three different main products, depending on the conditions.[11,12]

BzCl (5 eq) pyridine
15 min (no cooling) 70%

**3.10**

BzCl (4 eq) pyridine/$CHCl_3$
-10°C (1 h) 25°C (30 hr) 70%

**3.11**

BzCl (7 eq) pyridine 25°C(30 hr) 17%

**3.12**

**Scheme 3.8**

Finally, it should also be mentioned that monosaccharides with protection on all hydroxyl groups except 1-OH can be made by taking advantage of the fact that the 1-OH is a hemiacetal that can be converted into a methyl glycoside. The methyl glycoside can later be cleaved in the presence of other ethers by acidic hydrolysis (Scheme 3.9).[13]

Methyl glucoside

BnCl, KOH

$H_3O^+$

70% (2 steps)

3.66

Scheme 3.9

## SELECTIVE SUBSTITUTION OF HYDROXYL GROUPS

The different OH groups of the carbohydrate, when converted to a sulfonic ester or carboxylic ester, have roughly a leaving group ability in the following descending order: 1-OH >> 6-OH >> 4-OH ~ 3-OH > 2-OH for a hexopyranoside and 1-OH >> 5-OH >> 3-OH > 2-OH for a pentofuranoside. Substitution with nucleophiles in the 1-position is so easy that acetate often is a good enough leaving group. This is used in the important synthesis of **3.29** (Scheme 3.10).[14] For

D-Glucose

$Ac_2O$, $H^+$

$Br_2$, P

80-85%

3.29

Scheme 3.10

substitution with chloride in other positions, sulfuryl chloride can be used. This reagent converts alcohols into chlorosulfates, which, depending on structure, are readily substituted by liberated chloride ion. Unreacted chlorosulfates can easily be removed with iodide in methanol. When the 1-position is blocked, the 6-position is the first to be substituted. Later, some secondary chlorosulfates react. The reactivities of leaving groups at the secondary positions are, however, very small and are very dependent on configuration. The following two rules have been suggested by Richardson[15] for the substitution of sulfonate esters:

1. An axial electronegative substituent $\beta$ to the leaving group impedes substitution.
2. A transaxial electronegative substituent $\gamma$ to the leaving group impedes substitution.

These rules explain the observation that reaction of methyl-$\alpha$-D-glucopyranoside with sulfuryl chloride cleanly gives[16] dichloride **3.37** (Scheme 3.11) and not any

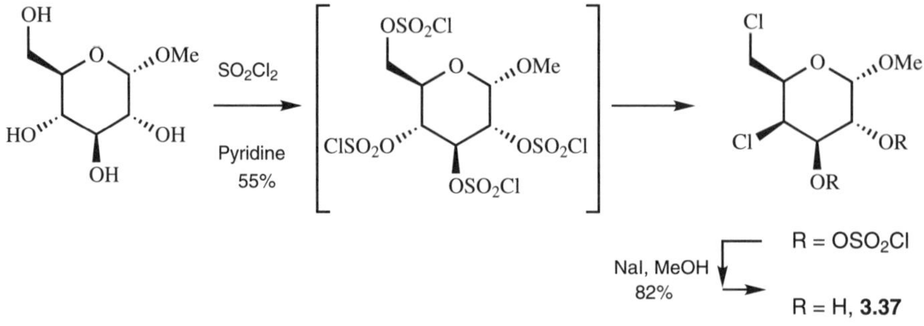

Scheme 3.11

3-substituted product. The axial methoxy group prevents substitution at C-3. This is even better illustrated in the reaction of methyl $\alpha$- and $\beta$-L-*arabino*-pyranoside with sulfuryl chloride[17] (Scheme 3.12). The $\alpha$-anomer gives a 3,4-dichloride, whereas the $\beta$-anomer only gives a 4-monochloride. Again the explanation is the impeding effect of an axial anomeric methoxy group on substitution in the 3-position.

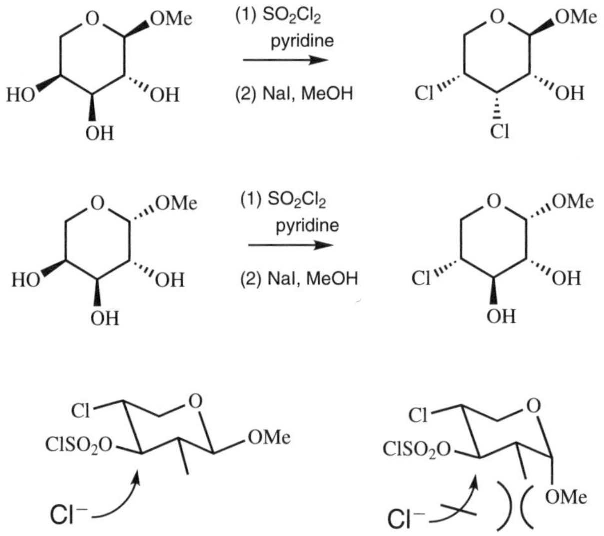

Scheme 3.12

## SELECTIVE SUBSTITUTION OF HYDROGEN

By radical bromination it is possible to selectively substitute certain hydrogens with bromine. These reactions can be very high yielding and useful (Scheme 3.13).[18,19] Generally C-5 can be selectively brominated in peracetylated sugars. In certain other cases (i.e., more electronegative glycosyl substituent), radical bromination occurs at C-1.

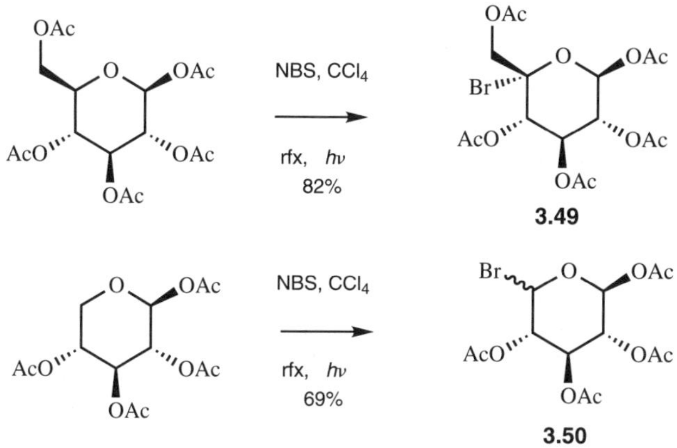

Scheme 3.13

## USE OF SELECTIVELY HALOGENATED CARBOHYDRATES

The selectively halogenated (or tosylated) carbohydrates that can be prepared by the methods previously described can be deoxygenated, eliminated, converted to epoxides, or substituted with nucleophiles. Nucleophilic substitution is however, often surprisingly difficult, as outlined in the rules by Richardson already given, and radical reactions are therefore more useful. For example, in the case of the deoxygenation of derivative **2.17**, Barton deoxygenation works perfectly,[20] while attempted nucleophilic substitution leads to rearrangement[21] (Scheme 3.14.) The expected substitution of the 4-chlorosulfonate is impeded by the $\gamma$-transaxial 2-ether.

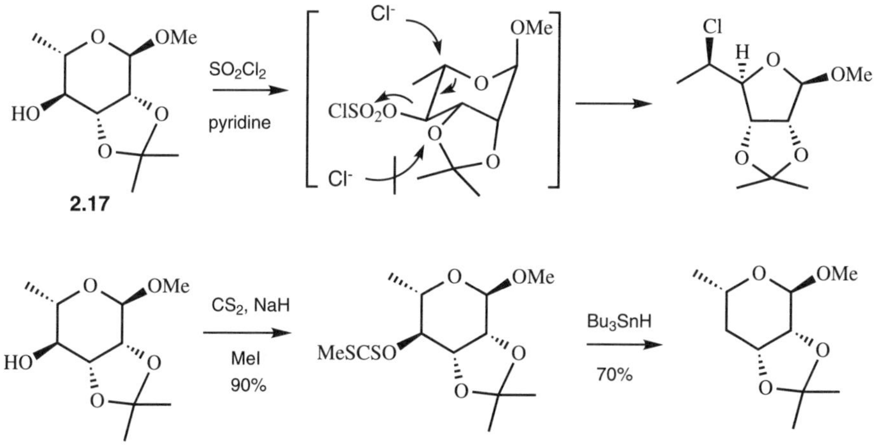

Scheme 3.14

## REFERENCES

1. F. Cramer, H. Otterbach, and H. Springbaum. *Chem. Ber.* **1959** *92:*384–391.
2. H. Pfander and M. Läderach. *Carbohydr. Res.* **1982** *99:*175–179.
3. A. H. Haines. *Adv. Carbohydr. Chem.* **1976** *33:*11–109.
4. J. Jary, K. Capek, and J. Kovar. *Collect. Czech. Chem. Commun.* **1964** *29:*930–937.
5. E. J. Reist, R. R. Spencer, D. F. Calkins, B. R. Baker, and L. Goodman. *J. Org. Chem.* **1965** *30:*2313–2317.
6. A. C. Richardson, J. M. Williams. *Tetrahedron* **1967** *23:*1641–1646.
7. H. Paulsen. *Angew. Chem. Int. Ed.* **1982** *21:*155–224.
8. S. Koto, N. Morishima, M. Owa, and S. Zen. *Carbohydr. Res.* **1984** *130:*73–83.
9. S. David and S. Hanessian. *Tetrahedron* **1985** *41:*643–663.
10. R. R. Conteras, J. P. Kamerling, J. Bres, and J. F. G. Vliegenthart. *Carbohydr. Res.* **1988** *179:*411–418.
11. P. Briegl and R. Schinle. *Chem. Ber.* **1934** *67:*127–130.
12. P. Briegl and R. Schinle. *Chem. Ber.* **1933** *66:*325–330.
13. C. P. J. Glaudemans and H. G. Fletcher, Jr. *Methods Carbohydr. Chem.* **1972** *6:*373–376.
14. R. U. Lemieux. *Methods Carohydr. Chem.* **1963** *2:*221–222.
15. A. C. Richardson. *Carbohydr. Res.* **1969** *10:*395–402.
16. H. J. Jennings and J. K. N. Jones. *Can. J. Chem.* **1963** *41:*1151–1159.
17. H. J. Jennings and J. K. N. Jones. *Can. J. Chem.* **1965** *43:*3018–3025.
18. R. Blattner and R. J. Ferrier. *J. Chem. Soc. Perkin 1* **1980** 1523–1527.
19. R. J. Ferrier, S. R. Haines, G. J. Gainsford, and E. J. Gabe. *J. Chem. Soc. Perkin 1* **1984** 1683–1687.
20. V. Pozsgay and A. Neszmelyi. *Carbohydr. Res.* **1980** *85:*143–150.
21. M. Bols and O. Ottosen. Unpublished results.

# 4

# OXIDATION PRODUCTS. ALDONIC ACIDS AND LACTONES

Oxidation of the aldehyde of a monosaccharide with bromine converts it into an aldonic acid[1] (Scheme 4.1), which at acidic pH can form a lactone, in most cases the $\gamma$-lactone. The equilibrium can easily be shifted completely to the right by removal of water by simple evaporation.[2] These carbohydrate lactones have some unique properties making it possible to turn them speedily into valuable building blocks, as described below. Furthermore, oxidative cleavage of a monosaccharide with oxygen and base give an aldonic acid/lactone with one carbon less.[3] This is a very useful transformation (Scheme 4.2). The aldonic acid/lactone with an ex-

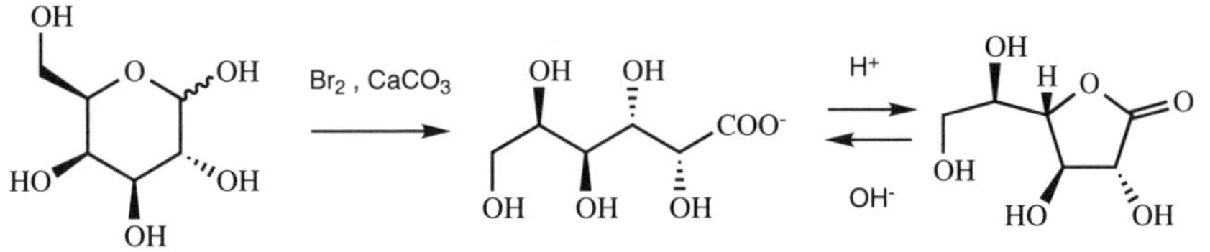

D-Galactose

$Br_2$, $CaCO_3$

$H^+$

$OH^-$

**Scheme 4.1**

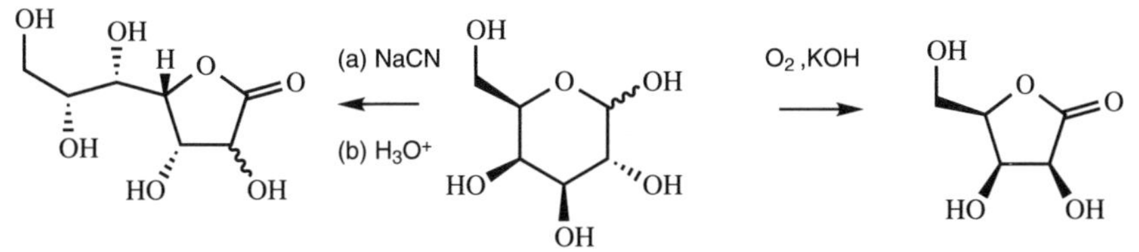

C7-lactone

(a) NaCN

(b) $H_3O^+$

D-Galactose

$O_2$, KOH

C5-lactone

**Scheme 4.2**

tra carbon can be prepared by reacting the monosaccharide with cyanide to a nitril followed by hydrolysis.[4] This is a less useful transformation because two stereoisomers are formed, but in singular cases one stereoisomer is particularly easy to isolate pure. Many sugar lactones are commerically available.

The carbohydrate lactone is so electronically different from the parent sugar that one can view it as if the sugar has been subjected to *umpolung*. In the lactone structure all the hydroxyl groups have different properties (Scheme 4.3). The 2-



**Scheme 4.3**

OH is easily substituted because it neighbors the ester, while the 3-OH readily undergoes $\beta$-elimination. The 4-OH is temporarily protected as a part of the lactone, while the 6-OH can be selectively protected or substituted because it is primary. The 5-OH differs by having no special properties. Therefore it is possible to do very selective reactions with these compounds. For instance, the 2- and 5-hydroxyl groups of ribonolactone can both be selectively substituted with bromide (Scheme 4.4). Mitsunobu conditions introduce the bromine at the less hindered primary position,[5] while HBr in acetic acid can be used to substitute the 2-OH with bromine. In the latter case a 2,3-acetoxonium ion is formed, which is selectively substituted next to the carbonyl group.[6]



**Scheme 4.4**

As earlier discussed, a problem with using carbohydrate derivatives as building blocks for synthesis is the presence of too many undesirable hydroxy groups. Therefore the HBr–acetic acid reaction of lactones is particularly useful because the bromide can be eliminated from the molecule in various ways to produce compounds with relatively simple structure. The reaction is quite general. Bromide is introduced at the 2-position if the 2,3-OH groups are cis, and at the primary position if the exocyclic chain contains at least two OH groups.[7] Accordingly, D-mannonolactone with HBr in acetic acid gives 2,6-dibromide **4.54** whereas D-galactonolactone gives a monobromide (Scheme 4.5). An important exception to the rule is D-gluconolactone, which gives dibromide **4.47** despite the fact that the 2,3-OH groups are in a trans relationship. The lactone ring probably opens in this case so that the reaction with formation of acetoxonium ions can take place on an acyclic form. Lactones with only four carbons react similarly[8] (Scheme 4.5).

D-Mannonolactone → HBr-HOAc, 80% → **4.54**

D-Galactonolactone → HBr-HOAc, 60-65% →

D-Gluconolactone → HBr-HOAc, 40-45% → **4.47**

HBr-HOAc, 73% → [COOH] → pH 3 → **4.19**

**Scheme 4.5**

Bromide can be displaced with hydrogen[9] or nucleophiles[10–12] (Scheme 4.6). In dibromolactones the large difference in the chemical surroundings of the two bromides makes it possible to discriminate between them. The bromine $\alpha$ to the ester reacts faster than the primary bromine. Two powerful reactions involve elimination of the 2,3-halohydrin. Reaction with hyrogensulfite converts a 2-bromolactone to the 2,3-unsaturated compound,[13,14] a butenolide (Scheme 4.6).

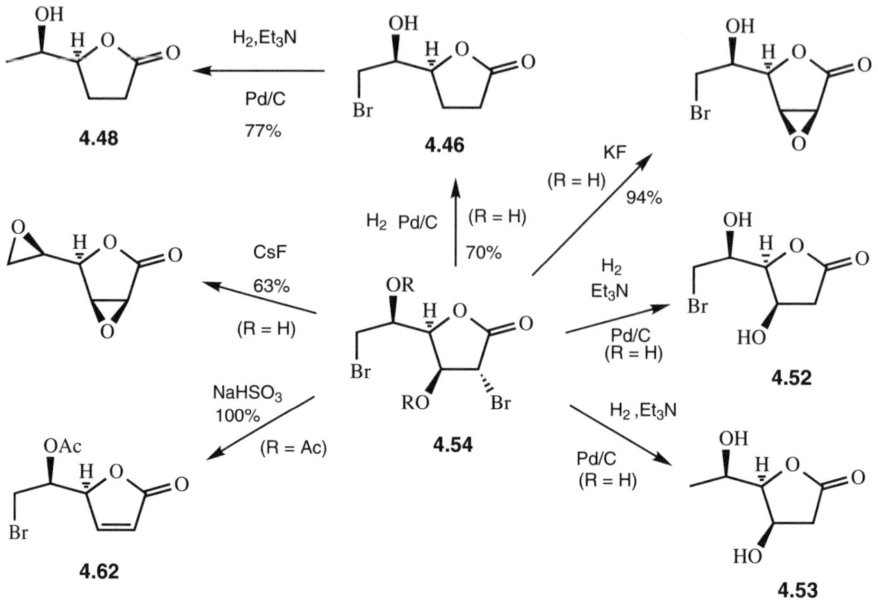

**Scheme 4.6**

Thus, a range of different butenolides can be prepared in only three steps from commercially available monosaccharides. Butenolides have particularly attracted interest because carbon functionalities can be introduced stereoselectively at C-2 and C-3.[15] The corresponding saturated lactones can similarly be prepared in one step from the halohydrin by hydrogenation with palladium on carbon without an acid acceptor,[16,17] a reaction discovered by Pedersen and Sonnichsen (Scheme 4.6).

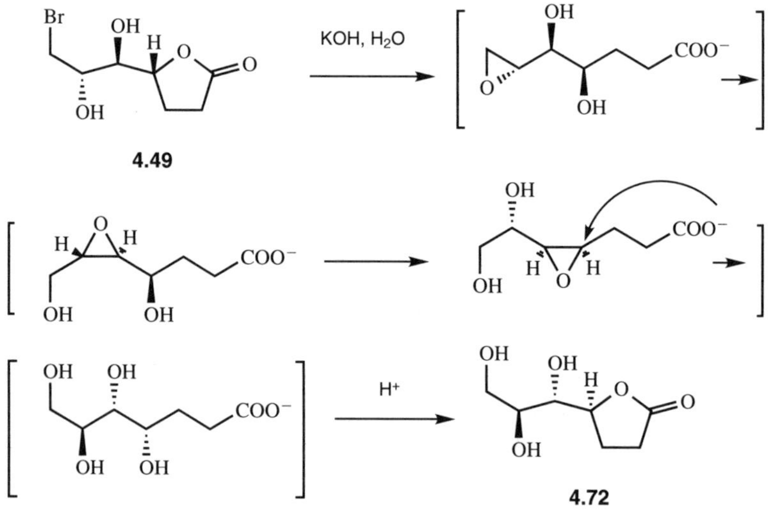

**Scheme 4.7**

The Pedersen–Sonnichsen reaction can be carried out in the presence of a primary bromide.

In certain cases, base-induced epoxide migration/hydrolysis of carbohydrate lactones are clean reactions and thus very useful because epimeric compounds

OH OH H O O OH HO OH
$Ac_2O$ $H^+$
OAc OAc H O O OAc AcO OAc
BzCl Pyridine
$H_2$,$Et_3$N Pd/C
OBz OBz H O O OBz OBz
**4.57**
$H_2$,Pd/C
OR OR H O O OR OR
R = Ac or Bz

**Scheme 4.8**

HO OH H O O HO OH
$Me_2C(OMe)_2$ TsOH, 79%
O O H O O O O
**4.78**
$H^+$, 79%
HO OH H O O O O
**4.81**

HO OH H O O HO OH
PhCHO, HCl 65%
OH H O O O Ph O OH
**4.79**

OH O O HO OH OH
$Me_2C(OMe)_2$ $H^+$
O O HO COOMe O O
**4.85**

**Scheme 4.9**

are obtained. The most elegant example is shown in Scheme 4.7, where inversion of all three chrial centers in one step is obtained.[18]

Another useful transformation of carbohydrate lactones is base-catalyzed elimination of an acylated 3-OH group to give the 2,3-unsaturated ester. Subsequent stereospecific hydrogenation gives a 3-deoxylactone, possibly with retention or reversal of the stereochemistry at C-2 (Scheme 4.8). Either acylation and elimination can be carried out in one step using benzoyl chloride/pyridine,[19,20] or elimination/hydrogenation can be done simultaneously on the lactone acetate.[21]

Selective protection of carbohydrate lactones with cyclic acetals is obviously a useful possibility (Scheme 4.9). As with monosaccharides, an acetonide in the exocyclic chain is more acid-labile than a ring-fused one.[22] Interestingly, acetonide formation by transacetalization can give rise to acyclic compound through methylester formation.[23]

## REFERENCES

1. J.W. Green. *Adv. Carbohydr. Chem. Biochem.* **1948** *3:*129–184.
2. H. S. Isbell and H. L. Frush. *Methods Carbohydr. Chem.* **1963** *2:*16–18.
3. W. J. Humphlett. *Carbohydr. Res.* **1967** *4:*157–164.
4. C. S. Hudson. *Adv. Carbohydr. Chem.* **1945** *1:*1–36.
5. I. Lundt. Personal communication.
6. K. Bock, I. Lundt, and C. Pedersen. *Carbohydr. Res.* **1981** *90:*17–26.
7. K. Bock, I. Lundt, and C. Pedersen. *Pure Appl. Chem.* **1978** *50:*1385–1400.
8. K. Bock, I. Lundt, and C. Pedersen. *Acta. Chem. Scand.* **1983** *B37:*341–344.
9. K. Bock, I. Lundt, and C. Pedersen. *Carbohydr. Res.* **1981** *90:*7–16.
10. M. Bols and I. Lundt. *Acta. Chem. Scand.* **1988** *B42:*67–74.
11. M. Bols and I. Lundt. *Acta. Chem. Scand.* **1990** *44:*252–256.
12. M. Bols, I. Lundt, and E. R. Ottosen. *Carbohydr. Res.* **1991** *222:*141–149.
13. J. A. J. M. Vekemans, G. A. M. Franken, C.W Dapperens, E. F. Godefroi, and G. J. F. Chittenden. *J. Org. Chem.* **1988** *53:*627–633.
14. J. A. J. M. Vekemans, C.W. Dapperens, R. Claessen, A. M. J. Koten, E. F. Godefroi, and G. J. F. Chittenden. *J. Org. Chem.* **1990** *55:*5336–5344.
15. K. Tomioka, T. Ishiguro, Y. Iitaka, and K. Koga. *Tetrahedron* **1984** *40:*1303–1312.
16. K. Bock, I. Lundt, C. Pedersen, and R. Sonnichsen. *Carbohydr. Res.* **1988** *174:*331–340.
17. I. Lundt and C. Pedersen. *Synthesis* **1986** 1052–1054.
18. K. Bock, I. Lundt, and C. Pedersen. *Carbohydr. Res.* **1988** *179:*87–96.
19. O. J. Varela, A. F. Cirelli, and R. M. de Lederkremer. *Carbohydr. Res.* **1979** *70:*27–35.
20. L. F. Sala, A. F. Cirelli, and R. M. de Lederkremer. *Carbohydr. Res.* **1980** *78:*61–66.
21. K. Bock, I. Lundt, and C. Pedersen. *Acta. Chem. Scand.* **1981** *B35:*155–162.
22. G.W. J. Fleet, N. G. Ramsden, and D. R. Witty. *Tetrahedron* **1989** *45:319–326.*
23. H. Regeling, E. de Rouville, and G. J. F. Chittenden. *Recl. Trav. Chim. Pays-Bas* **1987** *106:*461–464.

# 5

# REDUCTION PRODUCTS. CARBOHYDRATE POLYOLS

Reduction of a monosaccharide with sodium borohydride or many other reducing agents converts it to a polyol. Industrially this transformation is carried out by catalytic hydrogenation, and some of the polyols, D-glucitol (**1.19**) and D-mannitol (**1.20**), are very inexpensive. The others can readily be made in the laboratory by sodium borohydride reduction of the parent sugar[1] (Scheme 5.1). The polyols obtained from galactose, xylose, and ribose are achiral and thus uninteresting for our purpose, but those obtained from D- or L-arabinose are useful.

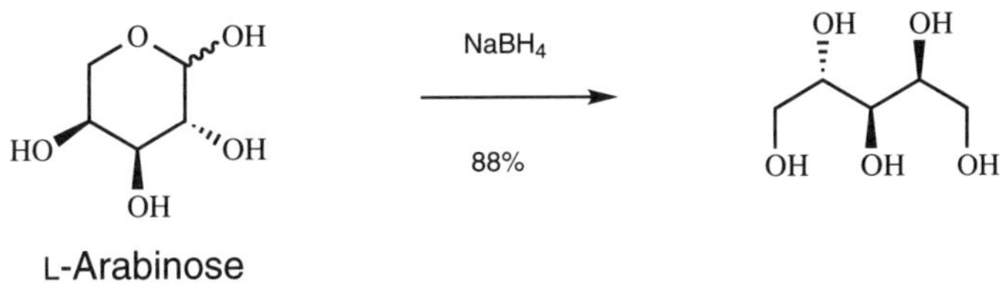


**Scheme 5.1**

## CYCLIC ACETALS

Partial protection of cyclic acetals is probably the best way of handling the many free hydroxyl groups of the carbohydrate polyols. Their chemistry crudely follows the general rules for cyclic acetal formation—namely, that aldehydes form six-membered rings and ketones form five-membered rings. Many useful partly protected derivatives can be made[2–4] (Schemes 5.2 and 5.3). The free hydroxy groups of these derivatives can be subjected to many transformations such as periodate cleavage or conversion to a leaving group, deoxygenation, elimination, or substitution. The use of these derivatives is further enhanced by the fact that an

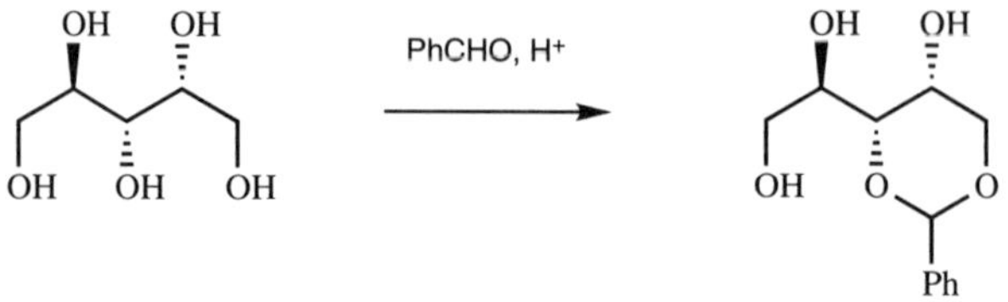

PhCHO, $H^+$

5.11

Scheme 5.2

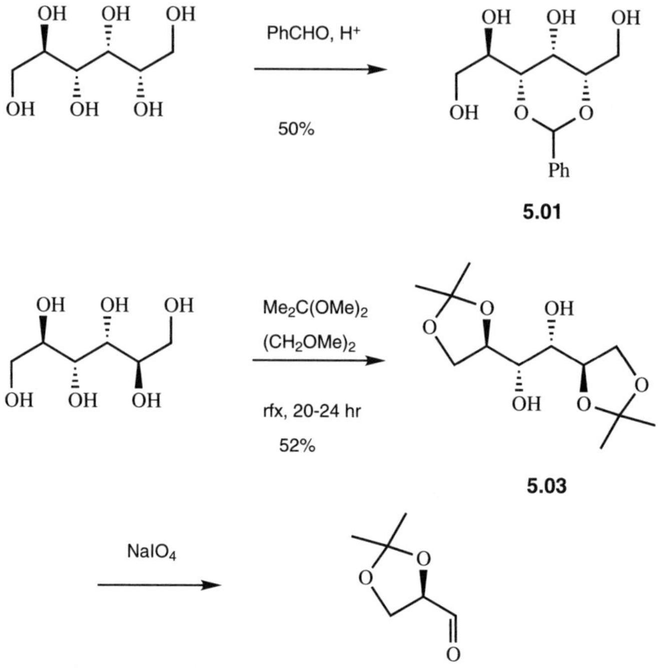

PhCHO, $H^+$

50%

5.01

$Me_2C(OMe)_2$

$(CH_2OMe)_2$

rfx, 20-24 hr

52%

5.03

$NaIO_4$

Scheme 5.3

acetonide of a diol, that include a primary alcohol, can be hydrolyzed selectively in the presence of one of a diol of two secondary alcohols.

## ANHYDROPOLYOLS

Treatment of carbohydrate polyols with acid leads to internal substitution, loss of water, and formation of five-membered rings, known as *anhydropolyols.* These can be useful building blocks for compounds containing tetrahydrofurans. D-Glucitol (**1.19**) by acid treatment gives anhydride **5.19** (Scheme 5.4).[5] Prolonged treatment gives dianhydride **1.21**, a compound that is commercially available at a low price. It is particularly interesting that discrimination is possible between the

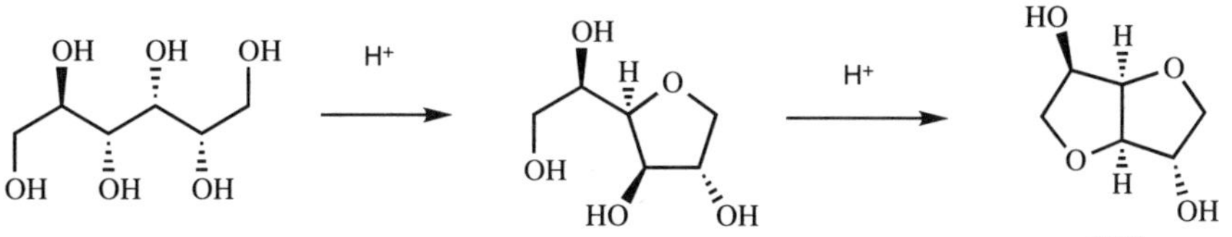

Scheme 5.4

two hydroxyl groups in **1.21**, because one is pseudoaxial and the other pseudoequatorial[6] (Scheme 5.5). Cleavage of the tetrahydrofurans to 1,4-chlorohydrins under mild conditions using $BCl_3$ further increases the usefulness of these compounds in synthesis.[7]

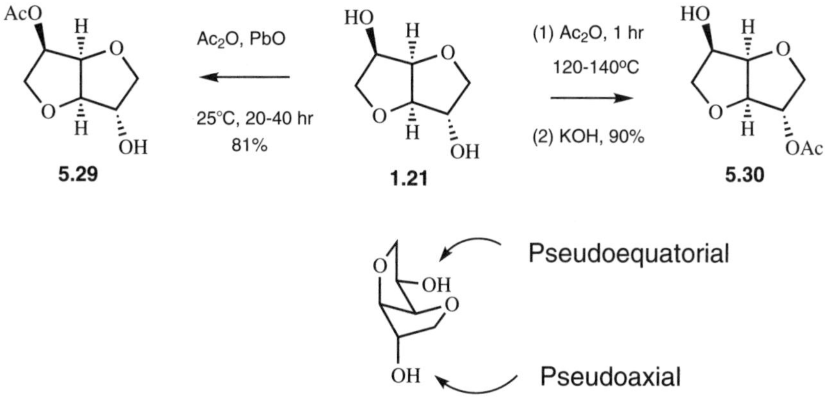

Scheme 5.5

## DITHIOACETALS

Reaction of monosaccharides with mercaptans under acid catalysis leads to formation of dithioacetals (Scheme 5.6).[8] These compounds have a chemistry very similar to that of the carbohydrate polyols because of the acyclic form, and therefore they are treated here (even though they are formally not reduction products of carbohydrates). The dithioacetal terminus obviously adds some more flexibility because it can be reconverted to aldehyde or desulfurized to a methyl group.

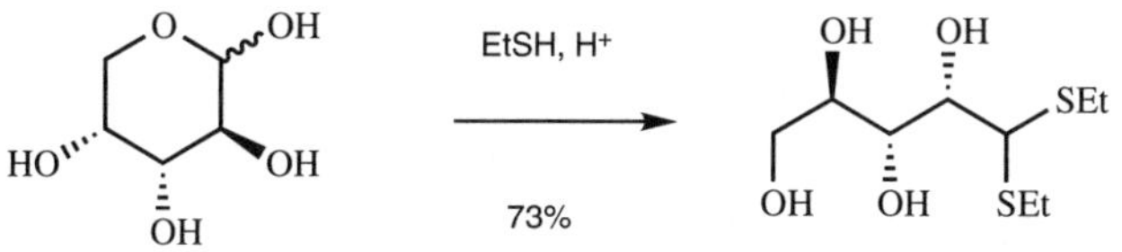

Scheme 5.6

The usual cyclic acetal chemistry is useful here as well (Scheme 5.7).[8–10]

$Me_2CO$, $H^+$

$CuSO_4$

**5.47**

$Me_2CO$, $H^+$

CuSO

76%

**5.48**

PhCHO

HCl/dioxane

79%

**5.41**

**Scheme 5.7**

## REFERENCES

1. M. Abdel-Akher, J. K. Hamilton, and F. Smith. *J. Am. Chem. Soc.* **1951** *73:*4691–4692.
2. E. Fisher. *Chem. Ber.* **1894** *27:*1524–1537.
3. L. von Vargha. *Chem. Ber.* **1935** *68:*18–24.
4. G. J. F. Chittenden. *Carbohydr. Res.* **1980** *87:*219–226.
5. S. Solzberg, R. M. Goepp, Jr., and W. Freudenberg. *J. Am. Chem. Soc.* **1946** *68:*919–921.
6. P. Stoss, P. Merrath, and G. Schlüter. *Synthesis* **1987** 174–176.
7. M. A. Bukhari, A. B. Foster, and J. M. Wobber. *Carbohydr. Res* **1966** *1:*474–481.
8. E. Pascu, S. M. Trister, and J.W. Green. *J. Am. Chem. Soc.* **1939** *61:*2444–2448.
9. E. J. Curtis and J. K. N. Jones. *Can. J. Chem.* **1960** *38:*890–895.
10. D. J. J. Potgieter and D. L. MacDonald. *J. Org. Chem.* **1961** *26:*3934–3938.

# 6

# 1,6-ANHYDRO SUGARS

1,6-Anhydro sugars, and especially 1,6-anhydroglucose (**6.01**), are a group of compounds of great potential use as building blocks due to their rigid structure and stability to many reagents. Their high crystallinity makes them easy to work with.

Furthermore 1,6-anhydroglucose (**6.01**, levoglucosan) is very readily available by pyrolysis of starch. Batches of several hundred grams of this compound can readily be made from a few kilograms of potato starch by pyrolysis in a vacuum. Detailed instructions for carrying out this operation can be found in the literature.[1–3]

Also important for the synthetic utility is that the anhydro bond can be hydrolyzed by strong acid (Scheme 6.1).[1]

Starch → Pyrolysis, 15-45% → **6.01** → $H_3O^+$ → D-Glucose

**Scheme 6.1**

Two other useful building blocks are 1,6-anhydrogalactose (**6.02**) and 1,6-anhydromannose (**6.03**). 1,6-anhydrogalactose (**6.02**) is formed together with **6.01** by pyrolysis of the inexpensive disaccharide lactose. **6.02** is separated from the mixture of **6.01** and **6.02** by conversion to the 3,4-isopropylidene derivative **6.04** and simple extraction.[4] Pure **6.02** can then be gained by hydrolysis (Scheme 6.2).

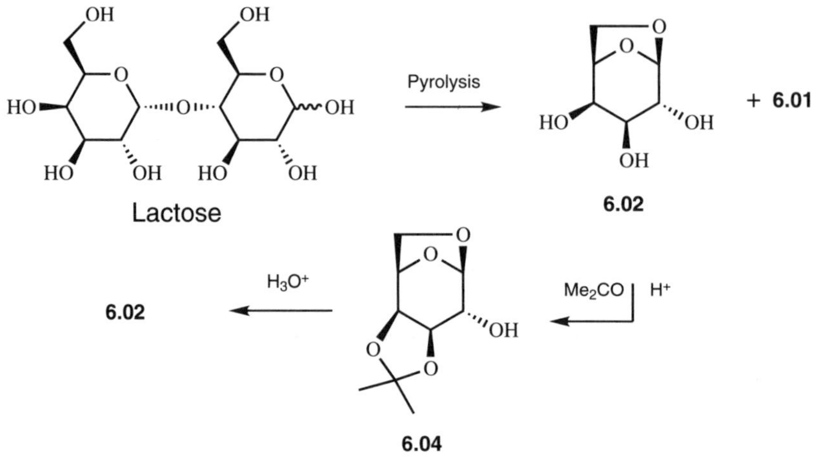

Lactose → Pyrolysis → **6.02** + **6.01**

**6.02** →($Me_2CO$, $H^+$) **6.04** →($H_3O^+$) **6.02**

**Scheme 6.2**

1,6-Anhydromannose (**6.03**) is obtained by pyrolysis of mannose. Again conversion to the 2,3-isopropylidene derivative **6.05** is used to separate the compound from other products (Scheme 6.3).[5,6] Of course the isopropylidene derivatives **6.04** and **6.05** are useful compounds themselves. They are used for preparing 2- or 4-substituted galactose or mannose derivatives, respectively.

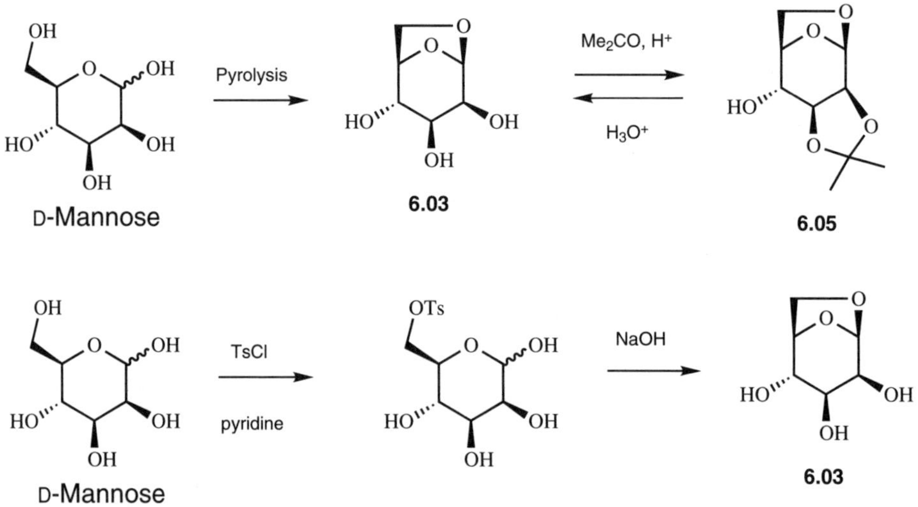

D-Mannose → Pyrolysis → **6.03** ⇌ ($Me_2CO$, $H^+$ / $H_3O^+$) **6.05**

D-Mannose → (TsCl, pyridine) → 6-O-Ts-mannose → (NaOH) → **6.03**

**Scheme 6.3**

**6.03** is probably more readily made by selective 6-*O*-tosylation of mannose followed by treatment with base (Scheme 6.3). In any case this procedure is experimentally simpler. **6.01** can also be prepared this way.[7]

Selective protection is a very useful method of preparing derivatives of 1,6-anhydro sugars because there generally is a considerable difference in reactivity of the various OH groups due to the locked $^1C_4$ conformation. Toward tosylation the relative reactivity of the various OH groups has been determined (Scheme 6.4).[1]

**Tosylation**

Decreasing reactivity

OH-2 equatorial
OH-3 equatorial
OH-2 axial
OH-4 axial
OH-4 equatorial
OH-3 axial

**Scheme 6.4**

Thus, tosylation of **6.01** gives the ditosylate **6.11** selectively,[8] while **6.03** can be tosylated selectively in the 2-position to give **6.15** (Scheme 6.5).[9] However, **6.02**

**6.01** → TsCl, Pyridine, 48-57% → **6.11** → NaOMe, 55% (from **6.01**) → **6.12**

**6.03** → TsCl, Pyridine, 48-57% → **6.15**

**Scheme 6.5**

cannot be tosylated selectively in the 4-position as might be expected. The ditosylate **6.11** selectively forms the 3,4-epoxide **6.12** when treated with base.[10] **6.12** is a useful building block for many syntheses. Epoxides such as **6.12** are valuable because they open regioselectively as a result of their locked conformation. A few transformations of **6.12** are shown in Scheme 6.6.[8,11–13]

Raney Ni $H_2$ 65-70%; BnOH, $H^+$ 62%; NaOMe 80%; $CH_2{=}CHCH_2MgCl$ 88%; NaOMe 89%

6.28 6.12 6.29 6.13 6.10

Scheme 6.6

Radical bromination is stereo- and regioselective on 1,6-anhydro sugars. Carbon 6 is invariably brominated. Benzoylated 1,6-anhydroglucose gives bromide **6.19** in 78% yield[14] by photobromination (Scheme 6.7).

$Br_2$ $h\nu$ 78%

6.19

Scheme 6.7

It must be mentioned that synthesis of some substituted 1,6-anhydro sugars is most efficiently carried out from other sugar derivatives. Iodo- and azidoderivatives **6.25** and **6.37** are obtained in high yield from glucal[15] (Scheme 6.8). Intramolecular iodocyclization catalyzed by formation of a tin ether leads to

$(Bu_3Sn)_2O$ $I_2$, 84%; $NaN_3$ DMF/$H_2O$ 81%

6.25 6.37

Scheme 6.8

iododerivative **6.25**. Reaction with sodium azide gives substitution with retention of configuration to azide **6.37**. The 2,3-epoxide is probably the intermediate.

A fourth very useful anhydro sugar is levoglucosenone **6.26**.[16] It is obtained by pyrolysis of acid-treated cellulose or paper in approximately 10% yield (Scheme 6.9).[17,18] The compound can also be prepared by pyrolysis of acid-treated

Cellulose or starch

(a) $H^+$ (b) pyrolysis 10%

**6.26**

**Scheme 6.9**

starch.[19] Despite the modest yield, it is no problem to make substantial amounts of this compound because of the extremely low price of the starting material. The relatively simple structure of **6.26** make it an attractive building block. The anhydro bridge causes addition reactions from the bottom face to be highly stereoselective. For example, 1,4-addition of dimethylcopper lithium to the enone leads to **6.55** in high yield (Scheme 6.10).[20] 1,2-Addition to the ketone, such as reduction,[21] is equally stereoselective, as is cycloaddition[22] to the double bond (Scheme 6.11). Many other transformations of **6.26** are known.[16]

**6.26** → $LiCuMe_2$, 86% → **6.55**

**Scheme 6.10**

**6.47** ← $LiAlH_4$, ether, 70% ← **6.26** → 98% → **6.41**

**Scheme 6.11**

## REFERENCES

1. M. Cerny and J. Stanek, Jr. *Adv. Carbohydr. Chem. Biochem.* **1977** *34:*23–177.
2. R. B. Ward. *Methods Carbohydr. Chem.* **1963** *2:*394–396.
3. G. Zemplen and A. Gerecs. *Chem. Ber.* **1931** *64:*1545–1554.
4. R. M. Hann and C. S. Hudson. *J. Am. Chem. Soc.* **1942** *64:*2435–2438.
5. K. Heyns, P. Köll, and H. Paulsen. *Chem. Ber.* **1971** *104:*830–836.
6. A. E. Knauf, R. M. Hann, and C. S. Hudson. *J. Am. Chem. Soc.* **1941** *63:*1447–1451.
7. M. A. Zottola, R. Alonso, G. D. Vite, and B. Fraser-Reid. *J. Org. Chem.* **1989** *54:*6123–6125.
8. M. Cerny, L. Kalvoda, and J. Pacak. *Collect. Czech. Chem. Commun.* **1968** *33:*1143–1156.
9. G. O. Aspinall and G. Zweifel. *J. Chem. Soc.* **1957** 2271–2278.
10. M. Cerny, V. Gut, and J. Pacak. *Collect. Czech. Chem. Commun.* **1961** *26:*2542–2550.
11. T. Trnka and M. Cerny. *Collect. Czech. Chem. Commun.* **1971** *36:*2216–2225.
12. A. G. Kelly and J. S. Roberts. *J. Chem. Soc. Chem. Commun.* **1980** 228–229.
13. M. Cerny and J. Pacak. *Collect. Czech. Chem. Commun.* **1962** *27:*94–105.
14. R. J. Ferier and R. H. Furneaux. *Aust. J. Chem.* **1980** *33:*1025–1036.
15. D. Tailler, J.-C. Jacquinet, A.-M. Noirot, and J.-M. Beau. *J. Chem. Soc. Perkin I* **1992** 3163–3164.
16. Z. J. Witczak. "Levoglucosenone and Levoglucosans. Chemistry and Applications," *Frontiers in Biomedicine and Biotechnology,* ATL Press, Mount Prospect, IL, vol. 2 **1994**.
17. F. Shafizadeh, R. H. Furneaux, and T. T. Stevenson. *Carbohydr. Res.* **1979** *71:*169–191.
18. R. H. Furneaux, J. M. Mason, and I. J. Miller. *J. Chem. Soc. Perkin I* **1984** 1923–1928.
19. F. Shadifazeh and P. P. S. Chin. *Carbohydr. Res.* **1974** *38:*177–187.
20. K. Mori, T. Chuman, and K. Kato. *Carbohydr. Res.* **1984** *192:*73–86.
21. K. Matsumoto, T. Ebata, K. Koseki, K. Okano, H. Kawakami, and H. Matsushita. *Carbohydr. Res.* **1993** *246:*345–352.
22. D. D. Ward and F. Shafizadeh. *Carbohydr. Res.* **1981** *95:*155–176.

# 7

# UNSATURATED SUGARS

Monosaccharides containing unsaturation are valuable synthons, particularly for making branched-chain compounds. A portion of this topic has recently been excellently reviewed.[1]

The most readily available unsaturated sugars are the 1,2-glycals. They serve both as useful building blocks in their own right and as precursors to most other unsaturated sugar derivatives. They are readily made from 1-halogeno sugars by reductive elimination with zinc (Scheme 7.1).[2] The reaction is general for making pyranoid glycals but fails for furanoid compounds. The former can thus be prepared from the parent sugar in two or three steps. Many addition reactions to these compounds are possible.

OAc O Br AcO OAc OAc **3.29** — Zn, HOAc, 75-80% → OAc O AcO OAc **7.02** — NaOMe, 73% → OH O HO OH **7.01**

**Scheme 7.1**

Reaction of the glycal with Lewis acid and an alcohol leads to a 2,3-unsaturated sugar by the so-called Ferrier rearrangement (Scheme 7.2).[3] This reaction is of broad scope.

A similar reaction, which substitutes the alcohol with *m*-chloroperbenzoic acid, leads to 2,3-unsaturated lactones[4] (Scheme 7.3). Unsaturated lactones can also be

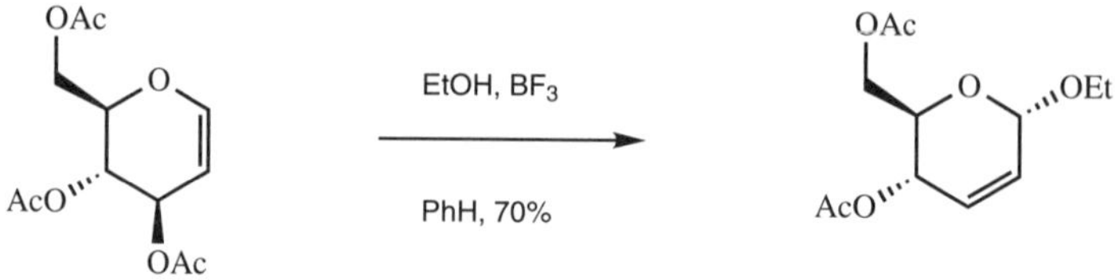

EtOH, $BF_3$

PhH, 70%

**7.02** **7.19**

**Scheme 7.2**

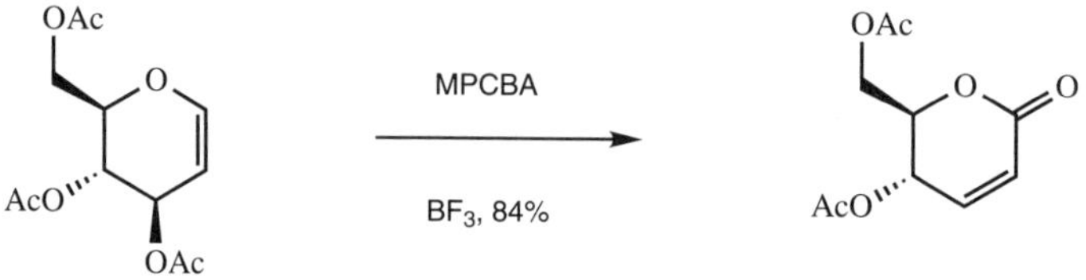

MPCBA

$BF_3$, 84%

**7.02** **7.46**

**Scheme 7.3**

made by elimination of acylated carbohydrate lactones (Chapter 4), but this procedure normally leads to 1,4-lactones and six-membered rings.

Treatment of glycals with mercury sulfate/sulfuric acid leads to acyclic $\alpha,\beta$-unsaturated aldehydes. This reaction can be performed in good yield on many glycals (Scheme 7.4).[5]

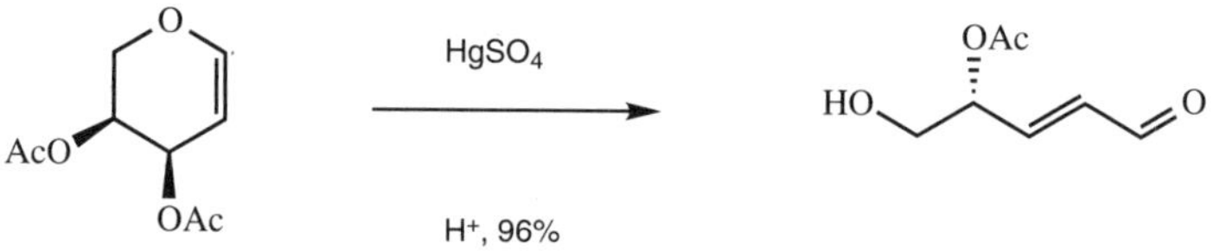

$HgSO_4$

$H^+$, 96%

**7.09** **7.30**

**Scheme 7.4**

Simple elimination of 1-halogeno sugars with a base gives 2-hydroxyglycals (Scheme 7.5).[6] These derivatives have a chemistry somewhat similar to that of the

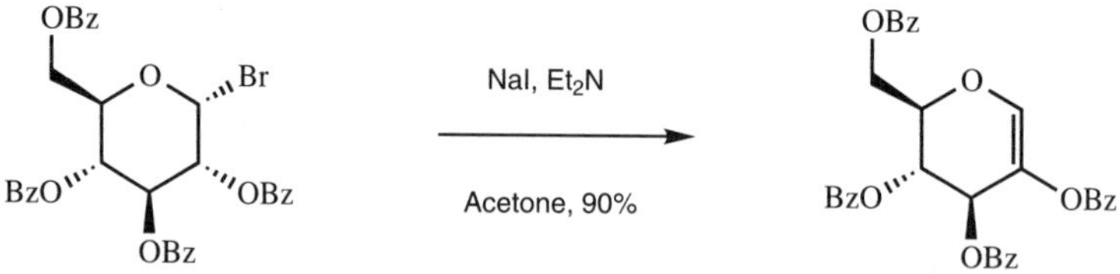

NaI, $Et_2N$

Acetone, 90%

**7.10**

**Scheme 7.5**

glycals; that is, they undergo Ferrier rearrangement and similar reactions as well as the various addition reactions.

Because addition reactions are greatly favored (both in terms of reactivity and regioselectivity) by the presence of a carbonyl group in the sugar, a great deal of effort has been done in preparing sugar enones. For example, the Ferrier product **7.19** can be deacetylated and oxidized at the allylic position to enone **7.26** (Scheme 7.6).[7] The reacetylated form of this compound has been subjected to a whole series of addition reactions (Scheme 7.7). These reactions include conjugate addition,[8] cycloadditions,[9,10] and radical addition.[11]

OAc, O, OEt, AcO, **7.19** — $Et_3N$, $H_2O$, 70% → OH, O, OEt, HO — $MnO_2$, 82% → OR, O, OEt, O; R = H **7.26**; $Ac_2O$ → R = Ac

Scheme 7.6

OR, O, OEt, O (R = Ac or Tr)
- $AlCl_3$, butadiene, 78% → OAc, O, OEt, O (cycloadduct)
- LiCuMe₂ → see below
- MeOH, $h\nu$, 50% → **7.35** (OAc, O, OEt, O, OH)
- $CH_2I_2$, Zn-Cu, 22% → **7.34** (OAc, O, OEt, O)

$LiCuMe_2$, 84% → **7.27** (OTr, O, OEt, O)

Scheme 7.7

A second type of enone can be obtained from the 2-hydroxy glycals. Ferrier rearrangement of glucal **7.11** with $Bu^tOH$ and $BF_3$ gives **7.22** (Scheme 7.8).[12] This compound, when treated with dimethylcopperlithium, leads to *in situ* deacetylation of the enol and $\beta$-elimination to the enone, which eventually undergoes conjugate addition.[12] In a similar manner, **7.22** under Wittig conditions gives diene **7.24**, again with the enone as intermediate (Scheme 7.9).[13] The enone can also be isolated.[14]

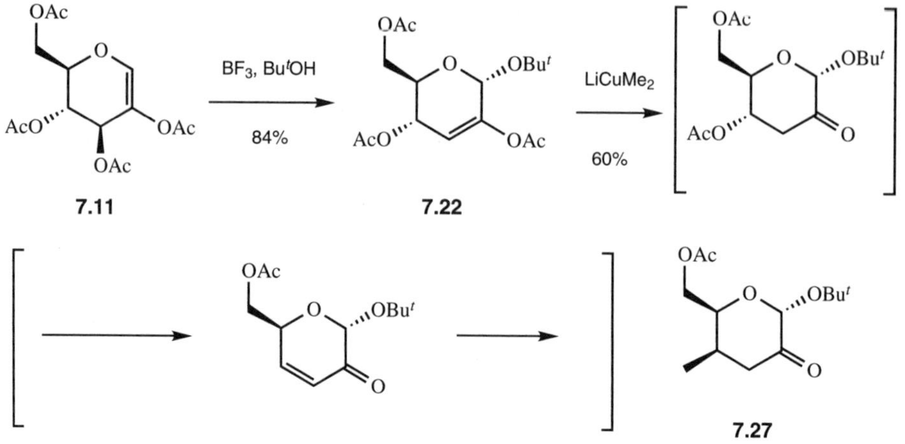

**Scheme 7.8**

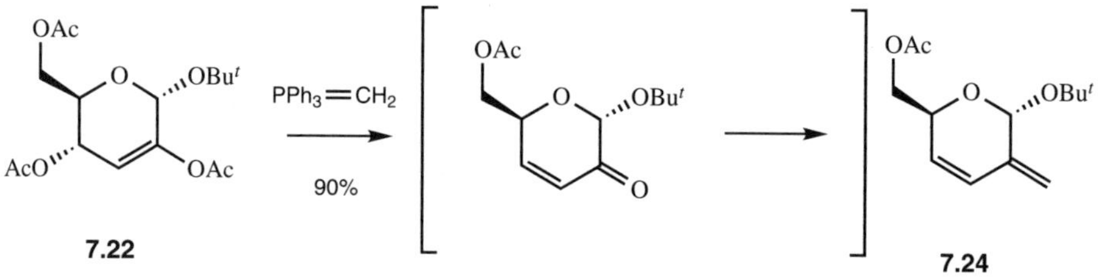

**Scheme 7.9**

A third type of enone is **7.53**. This compound can be obtained by allylic oxidation of glucal (Scheme 7.10).[15] Conjugate addition can then be carried out, giving C-glycoside-type structures.[16]

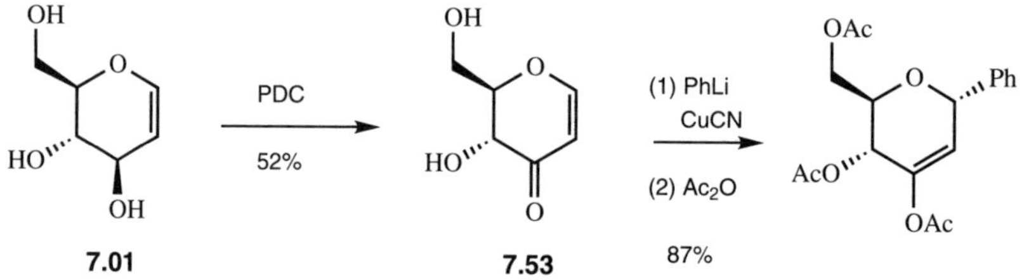

**Scheme 7.10**

## REFERENCES

1. F.W. Lichtenthaler. *Modern Synthetic Methods,* vol. 6. VCHA, Basel, **1992** pp. 273–376.
2. W. Roth and W. Pigman. *Methods Carbohydr. Chem.* **1963** *2:*405–408.

3. R. J. Ferrier and N. Prasad. *J. Chem. Soc. (C)* **1969** 570–575.
4. F.W. Lichtenthaler, S. Rönninger, and P. Jarglis. *Liebigs Ann. Chem.* **1989** 1153–1161.
5. J. Wengel, J. Lau, and E. B. Pedersen. *Synthesis* **1989** 829–832.
6. M. G. Blair. *Methods Carbohydr. Chem.* **1963** *2:*411–414.
7. B. Fraser-Reid, A. McLean, E.W. Usherwood, and M. Yunker. *Can. J. Chem.* **1970** *48:*2877–2884.
8. M. Yunker, D. E. Plaumann, and B. Fraser-Reid. *Can. J. Chem.* **1977** *55:*4002–4009.
9. B. Fraser-Reid and B. J. Carthy. *Can J. Chem.* **1972** *50:*2928–2934.
10. J. L. Primeau, R. C. Anderson, and B. Fraser-Reid. *J. Chem. Soc. Chem. Commun.* **1980** 6–8.
11. B. Fraser-Reid, N. L. Holder, and M. B. Yunker. *J. Chem. Soc. Chem. Commun.* **1972** 1286–1287.
12. S. Hanessian, P. C. Tyler, and Y. Chapleur. *Tetrahedron Lett.* **1981** *22:*4583–4586.
13. S. Hanessian, P. C. Tyler, G. Demailly, and Y. Chapleur. *J. Am. Chem. Soc.* **1981** *103:*6243–6246.
14. S. Hanessian, A.-M. Faucher, and S. Leger. *Tetrahedron* **1990** *46:*231–243.
15. S. Czernecki, K. Vijayakumaran, and G. Ville. *J. Org. Chem.* **1986** *51:*5472–5475.
16. V. Bellosta and S. Czernecki. *Carbohydr. Res.* **1987** *171:*279–288.

# 8

# PRODUCTS OF BASE TREATMENT

Reducing sugars undergo a complicated series of reactions when treated with base starting with enolization, epimerization, and rearrangement of the carbonyl group. The final result is normally a complicated mixture, but in two cases interesting and useful compounds are obtained by such reactions.

Treatment of D-fructose with aqueous calcium hydroxide at 25°C leads to a 9% yield of glucosaccharinic acid lactone (**8.01**).[1] Despite the low yield, the reaction is synthetically useful because of **8.02**'s interesting structure, because the starting materials are extremely inexpensive and because the reaction is easy to perform. However, it requires considerable planning since the optimal reaction time is 6–8 weeks. **8.01** is formed by rearrangement of the keto group in fructose to the 3-position (via enolization), along with $\beta$-elimination of the 1-OH group to form a diketone that undergoes benzilic acid rearrangement (Scheme 8.1).

In a similar manner, treatment of lactose with calcium hydroxide leads to isosaccharinic acid lactone (**8.10** Scheme 8.2).[2] In that case the carbonyl group rearranges to the 2-position by enolization and the galactosyl unit is eliminated to give a diketone that undergoes benzilic acid rearrangement.

Gluco- and isosaccharinic acid lactones (**8.01** and **8.10**) are useful carbon-branched compounds and have found considerable synthetic use.[3] Some of their chemistry is shown in Schemes 8.3 and 8.4. Because they have vicinal diols, selective manipulation of the hydroxy groups via formation of acetonides is possible.[4,5] Reduction of the lactone moiety in either acetonide **8.02** or **8.11** leads to acyclic compounds with vicinal diols at one terminus. These can be cleaved to the aldehydes **8.06**[6] and **8.13**,[7] respectively. It is particularly noteworthy that the 5-OH in **8.10** can be silylated selectively to give **8.17** without silylation of the other primary alcohol.[3]

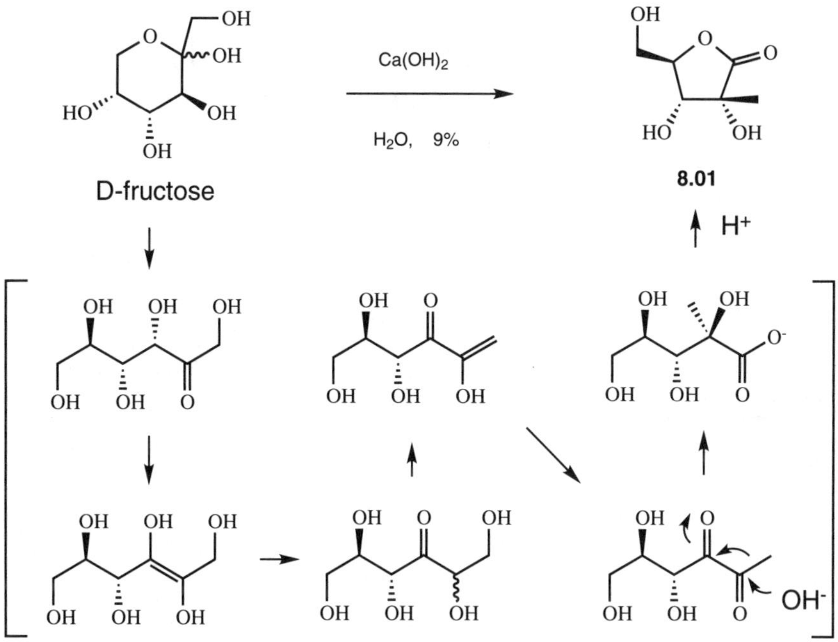

D-fructose

$Ca(OH)_2$

$H_2O$, 9%

**8.01**

$H^+$

$OH^-$

**Scheme 8.1**

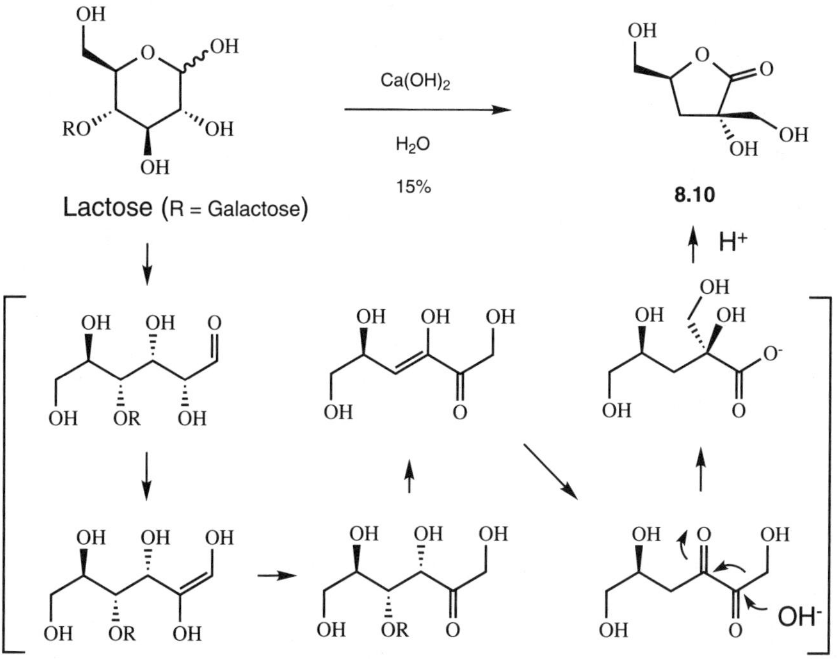

Lactose (R = Galactose)

$Ca(OH)_2$

$H_2O$

15%

**8.10**

$H^+$

$OH^-$

**Scheme 8.2**

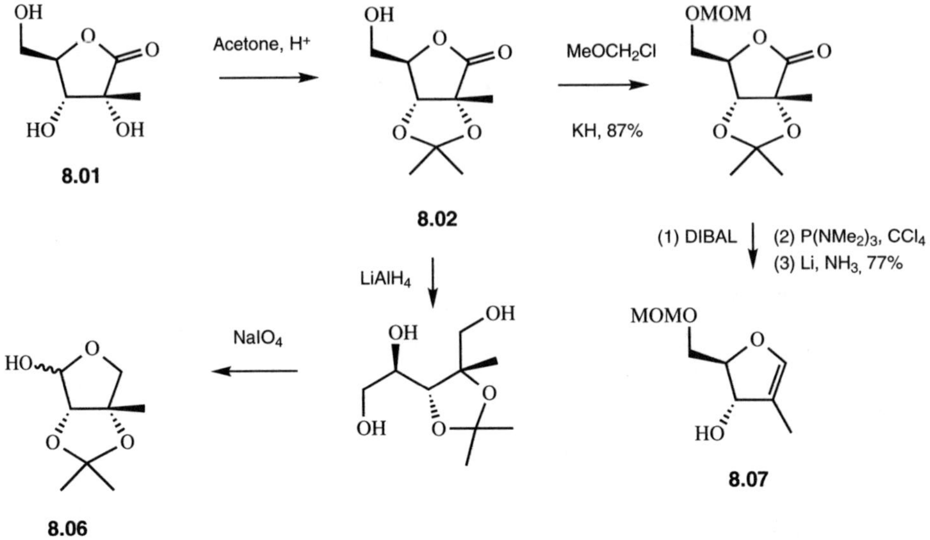

Scheme 8.3

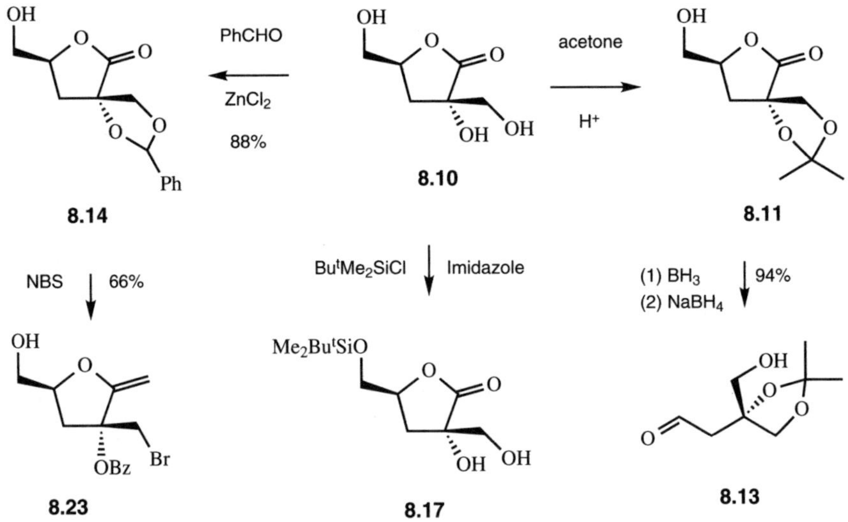

Scheme 8.4

## THE AMADORI REARRANGEMENT

The reaction of reducing sugars with amines constitutes the Maillard reaction, one of the most complex reactions known. However, the first part of the Maillard reaction, called the *Amadori rearrangement,* can be synthetically useful. Thus, D-glucose reacts with secondary amines to 1-amino-1-deoxy fructose derivatives in high yield (Scheme 8.5).[8] The reaction goes via the enamine and formation of the 2-ketone from the enol.

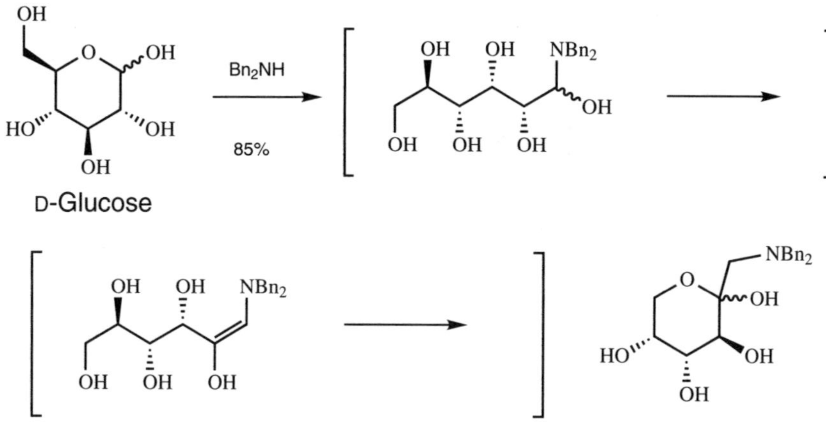

D-Glucose $\xrightarrow[85\%]{Bn_2NH}$ 8.19

Scheme 8.5

## ALDOL CONDENSATIONS WITH FORMALDEHYDE

Because reducing sugars degrade rapidly in base, base-catalyzed aldol-condensation reactions of these compounds are not expected to be very feasible. However, base-catalyzed aldol condensation between formaldehyde and reducing sugars containing an $\alpha,\beta$-acetonide or similar functionality that cannot epimerize is a very successful way of obtaining branched derivatives. Some examples are shown in Scheme 8.6.[9,10] When the reaction is carried out with a weak base, such as

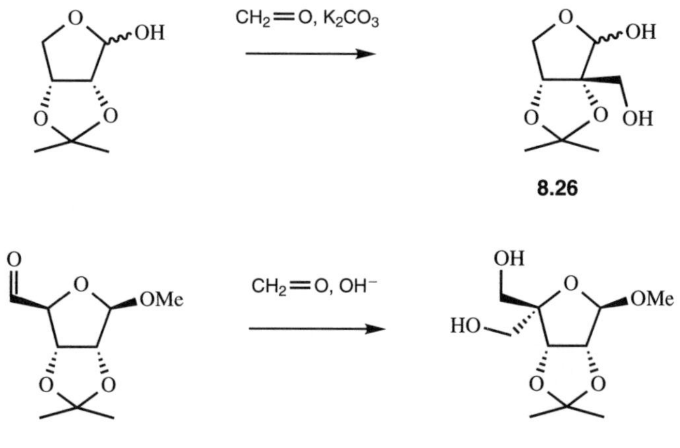

$CH_2{=}O$, $K_2CO_3$ → 8.26

$CH_2{=}O$, $OH^-$ → 8.27

Scheme 8.6

potassium carbonate, only aldol condensation occurs. However, when strong base is employed, Cannizzaro reaction between excess formaldehyde and the aldol occurs and a reduced product is obtained.

## REFERENCES


1. R. L. Whistler and J. N. Bemiller. *Methods Carbohydr. Chem.* **1963** *2:*484–485.
2. R. L. Whistler and J. N. Bemiller. *Methods Carbohydr. Chem.* **1963** *2:*447–479.
3. C. Monneret and J.-C. Florent. *Synlett* **1994** 305–318.
4. R. E. Ireland, R. C. Andersen, R. Baboud, B. J. Fitzsimmons, G. J. McGarvey, S. Thaisrivongs, and C. S. Wilcox. *J. Am. Chem. Soc.* **1983** *105:*1988–2006.
5. S. Hanessian and R. Roy. *Tetrahedron Lett.* **1981** *22:*1005–1008.
6. K. Ando, T. Yamada, Y. Takaishi, and M. Shibuya. *Heterocycles* **1989** *29:*1023–1027.
7. F. Bennani, J.-C. Florent, M. Koch, and C. Monneret. *Tetrahedron* **1984** *40:*4669–4676.
8. J. E. Hodge and B. E. Fisher. *Methods Carbohydr. Chem.* **1963** *2:*99–107.
9. P.-T. Ho. *Can. J. Chem.* **1980** *58:*858–860.
10. G. H. Jones, M. Taniguchi, D. Tegg, J. G. Moffatt. *J. Org. Chem.* **1979** *44:*1309–1317.

# 9

# PRODUCTS OF ACID TREATMENT

Treatment of reducing sugars with strong acid results in a number of elimination reactions taking place and final formation of furan products. Thus, treatment of a hexose with acid leads to 2-carboxaldehyde-5-hydroxymethylfurane (Scheme 9.1), an achiral compound which is uninteresting for the synthesis of chiral compounds. However, treatment of glucal with acid gives as initial product chiral dihydroxyethylfurane **9.01** (Scheme 9.2),[1] a useful building block. The enantiomer

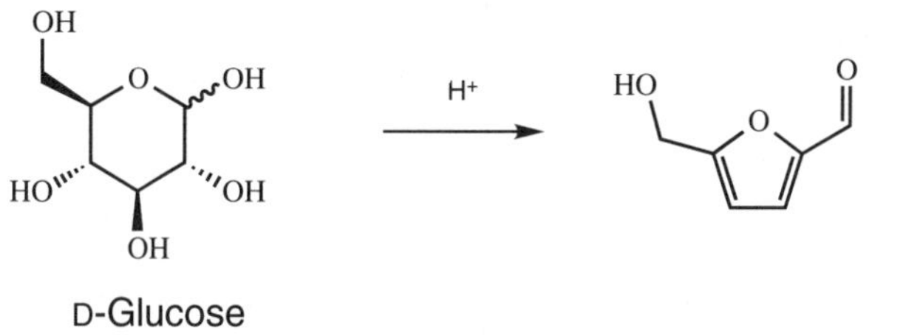

D-Glucose

$H^+$

**Scheme 9.1**

$H^+$, $HgSO_4$

81 %

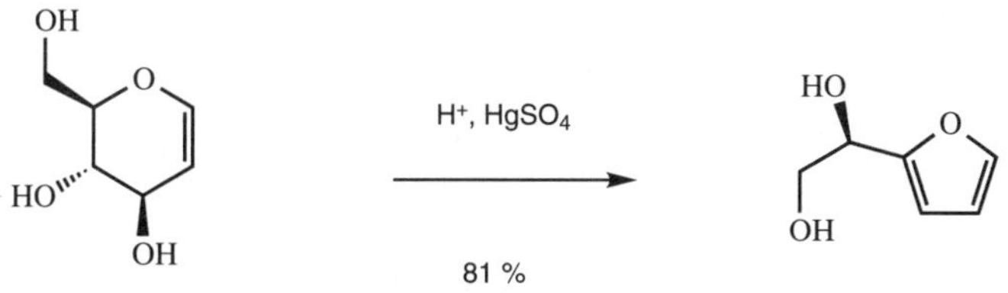

**9.01**

**Scheme 9.2**

of **9.01** can be prepared from **9.01** by Mitsunobu reaction.[2] Useful chiral products can also be obtained by the reaction between acetylated halo sugars and strong Lewis acids. The strong Lewis acid extracts the halogen of the 1-position to give an oxocarbenium ion (Scheme 9.3). In the case of glucose derivative **3.30**, a series of acetoxonium ion rearrangements results in formation of idose derivative **9.02** where three chiral centers have been inverted.[3] This constitutes an extremely easy way of preparing a sugar with a configuration that is otherwise difficult to obtain.

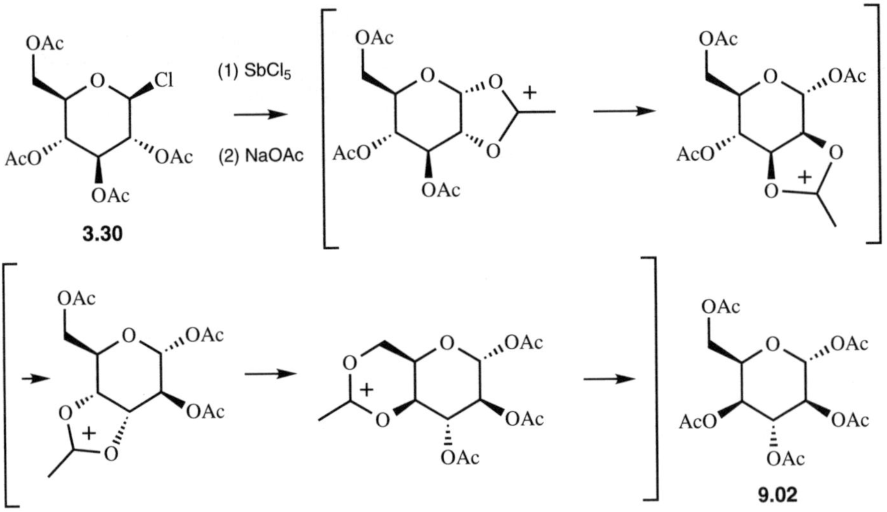

**Scheme 9.3**

## REFERENCES

1. F. Gonzales, S. Lesage, and A. S. Perlin. *Carbohydr. Res.* **1975** *42*:267–274.
2. F. M. Hauser, S. R. Ellenberger, and W. P. Ellenberger. *Tetrahedron Lett.* **1988** *29*:4939–4942.
3. H. Paulsen. *Methods Carbohydr. Chem.* **1972** *6*:142–148.

# 10

# DISACCHARIDES

Some of the least expensive available carbohydrates are a number of disaccharides. They have unfortunately found little synthetic use, and almost no employment outside the carbohydrate field. This is not surprising because the many functionalities make them difficult to handle in an efficient manner. However, the low price and little exploited chemistry of the disaccharides make them a potentially interesting field for creative chemistry. A beautiful example of the use

Sucrose

(a) $PPh_3$, $CCl_4$, 65-75%

(b) $NaN_3$, DMF, 81%

$H^+$ 62-64%

$H_2$, Pd/C, 78%

Glucose isomerase, 8-10%

**Scheme 10.1**

of sucrose for the synthesis of the natural product 1-deoxynoijirimycin has been carried out by de Raadt and Stütz (Scheme 10.1).[1] This synthesis takes advantage of the unreactiveness of the 1′-position to nucleophilic substitution (because it is next to the acetal function). Therefore the Mitsunobu-type reaction of sucrose with $CCl_4$ and $PPh_3$ allows selective substitution of two of the three primary hydroxy groups. After exchange of the chlorides to azides by substitution with $NaN_3$, hydrolysis of the glycosidic bond gives 6-azido-glucose and 6-azido-fructose. These two compounds can be separated and the enzyme glucose isomerase employed to convert 6-azido-glucose into 6-azido-fructose. Eventually 6-azido-fructose is subjected to hydrogenation, which results in intramolecular stereoselective reductive amination giving 1-deoxynoijirimycin.

Most disaccharide chemistry has been carried out on maltose due to its homogeneous nature consisting of two glucoses. Generally, much of the chemistry that can be used on monosaccharides is equally applicable to disaccharides. Thus, maltose can be converted into the glycosyl bromide[2] and hence the glycal **10.19**[3]

Maltose

a) $Ac_2O$
b) PhOH, $H^+$

(a) $Ac_2O$
(b) HBr-HOAc

(a) KOH
(b) $Ac_2O$

DBU

Zn, HOAc

**10.11**

**10.19**

**Scheme 10.2**

or hydroxyglycal or, by means of the 1-phenylglycoside, into the 1,6-anhydro sugar, **10.11** (Scheme 10.2).[4] Similar chemistry can be carried out on cellubiose and lactose.

It is possible to take advantage of the different steric environment of the two primary hydroxy groups. Maltose can be selectively tritylated or tosylated on the 6′-position due to more steric hindrance on the 6-position (Scheme 10.3).[5] On the

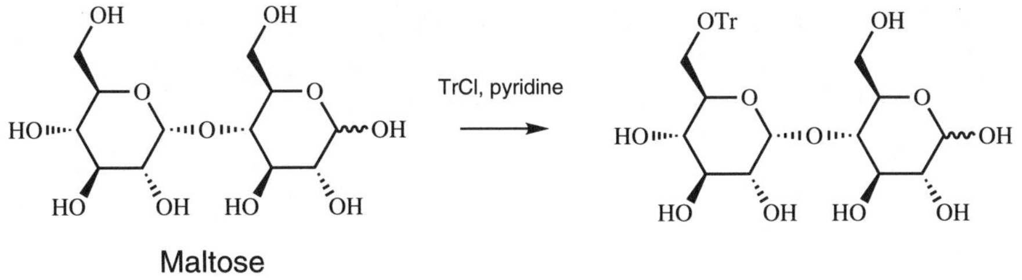

Maltose

TrCl, pyridine

**Scheme 10.3**

other hand, almost complete acetylation or benzoylation of maltose leaves the 3-OH, the least reactive OH group, free (Scheme 10.4).[6]

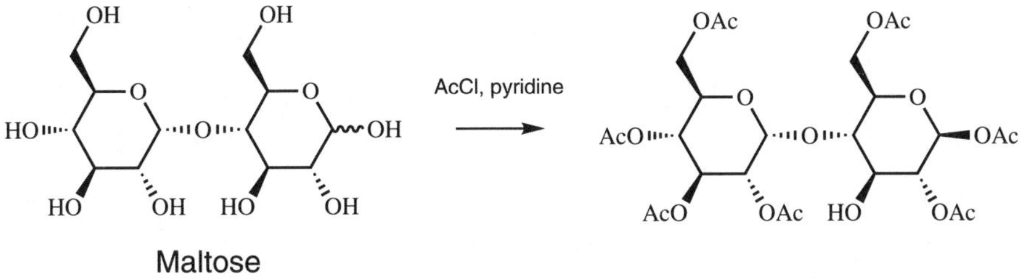

Maltose

AcCl, pyridine

**Scheme 10.4**

An interesting use of disaccharides is in reactions where the corresponding monosaccharide chemistry leads to achiral material. Thus, acid treatment of isomaltulose gives the interesting furane **10.25** (Scheme 10.5) having the glucose moiety as chiral auxiliary.[7]

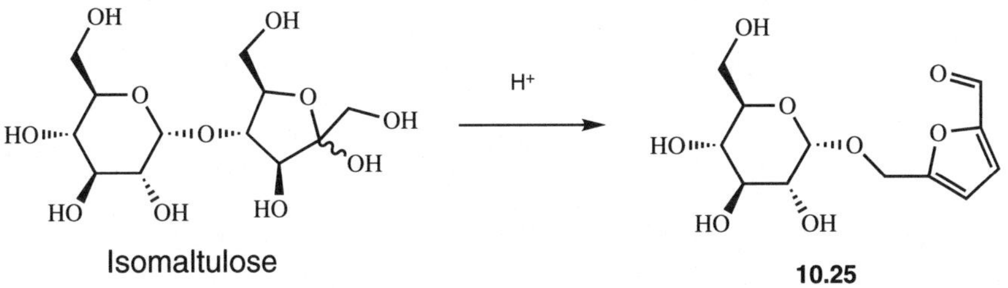

Isomaltulose

$H^+$

**10.25**

**Scheme 10.5**

## REFERENCES

1. A. de Raadt and A. E. Stütz. *Tetrahedron Lett.* **1992** 189–192.
2. D. H. Brauns. *J. Am. Chem. Soc.* **1929** *51:*1820–1831.
3. W. N. Haworth, E. L. Hirst, and R. J. W. Reynolds. *J. Chem. Soc.* **1934** 302–303.
4. L. Asp and B. Lindberg. *Acta Chem. Scand.* **1952** *6:*941–946.
5. K. Koizumi and T. Utamura. *Carbohydr. Res.* **1974** *33:*127–134.
6. W. E. Dick, B. G. Baker, and J. E. Hodge. *Carbohydr. Res.* **1968** *6:*52–62.
7. F. W. Lichtenthaler, D. Martin, T. A. Weber, and H. M. Schiweck. European Patent Application EP 426,176, *Chem. Abstract* **1991** *115:*92826t.

# 11

# MISCELLANEOUS CARBOHYDRATE PRODUCTS

Because it contains an amino group, glucosamine has some chemistry distinct from that of other sugars. Diazotation of glucosamine results in intramolecular substitution of an amino group and formation of a tetrahydrofurane (Scheme 11.1).[1]

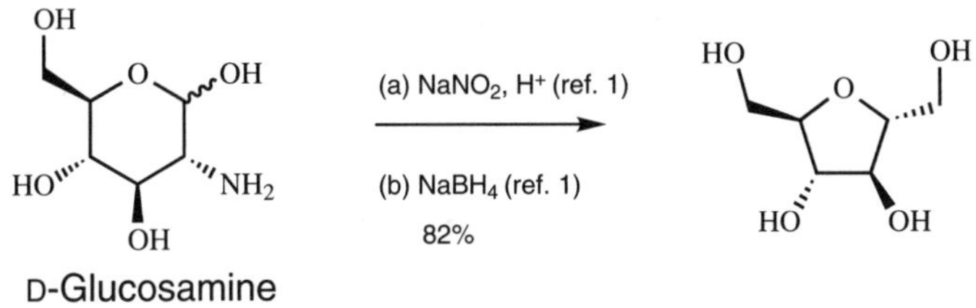

D-Glucosamine

(a) $NaNO_2$, $H^+$ (ref. 1)

(b) $NaBH_4$ (ref. 1)

82%

**Scheme 11.1**

An unusual method of preparing nitro sugars has been developed by Baer.[2] It involves periodate cleavage of a methyl glycoside to a dialdehyde, and then reacting with base and nitromethane. The reaction is quite stereoselective in favor of the product with exclusively equatorial substituents (Scheme 11.2).

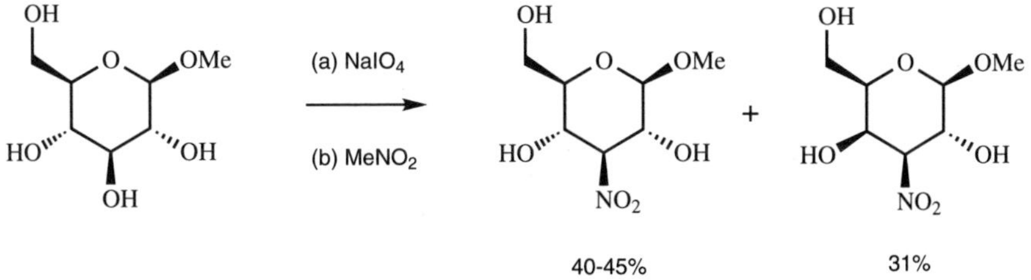

(a) $NaIO_4$

(b) $MeNO_2$

40-45%

31%

**Scheme 11.2**

## REFERENCES

1. D. Horton and K. D. Phillips. *Methods Carbohydr. Chem.* **1976** *7*:68–70.
2. H. Baer. *Methods Carbohydr. Chem.* **1972** *6*:245–249.

# COMPENDIUM OF BUILDING BLOCKS

EXAMPLE:

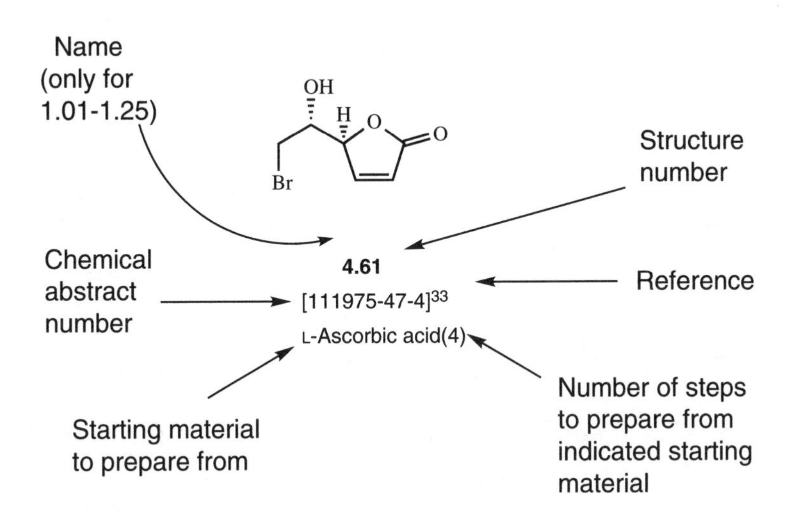


ABBREVIATIONS

| | |
|---|---|
| Dianhydroglucitol | 1,4:3,6-Dianhydro-D-glucitol |
| Dianh. glucitol | 1,4:3,6-Dianhydro-D-glucitol |
| Galactose | D-Galactose |
| Gluconolactone | D-Glucono-1,5-lactone |
| Gluc. lactone | D-Glucono-1,5-lactone |
| Glucuronolactone | D-Glucofuranuronic acid 3,6-lactone |
| Glucu. lactone | D-Glucofuranuronic acid 3,6-lactone |

| | |
|---|---|
| Heptonolactone | D-Glycero-D-gulo-heptono-1,4-lactone |
| Hept. lactone | D-Glycero-D-gulo-heptono-1,4-lactone |
| Methyl glucoside | Methyl $\alpha$-D-glucopyranoside |
| Me Glucoside | Methyl $\alpha$-D-glucopyranoside |

## COMMERCIAL MATERIALS

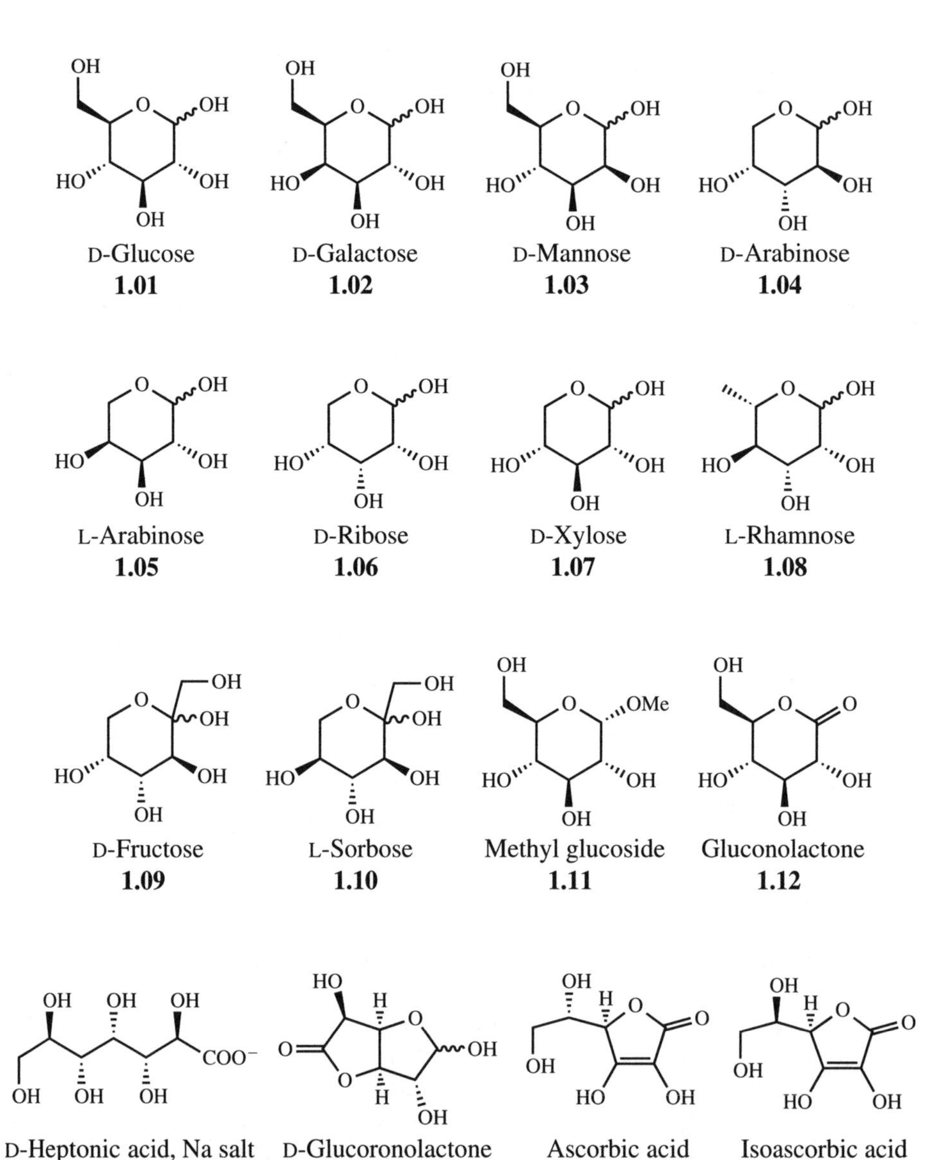

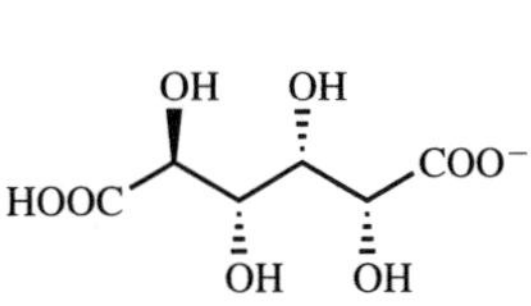

D-Glucaric acid
**1.17**

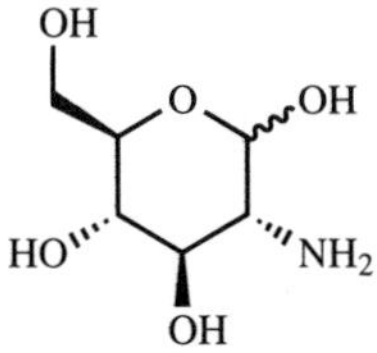

D-Glucosamine
**1.18**

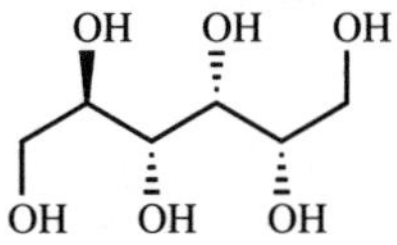

D-Glucitol
**1.19**

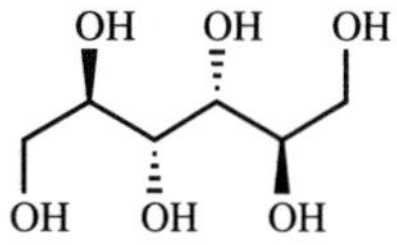

D-Mannitol
**1.20**

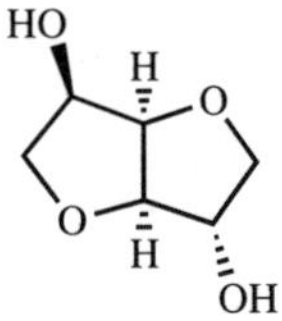

Dianhydroglucitol
**1.21**

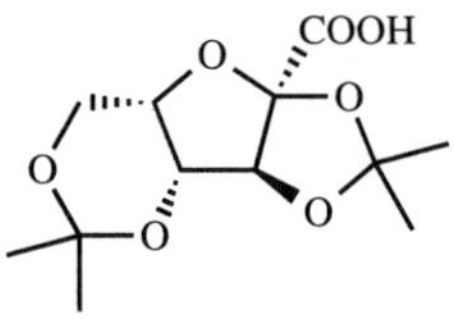

Diisopropylidene-2-ketogulonic acid
**1.22**

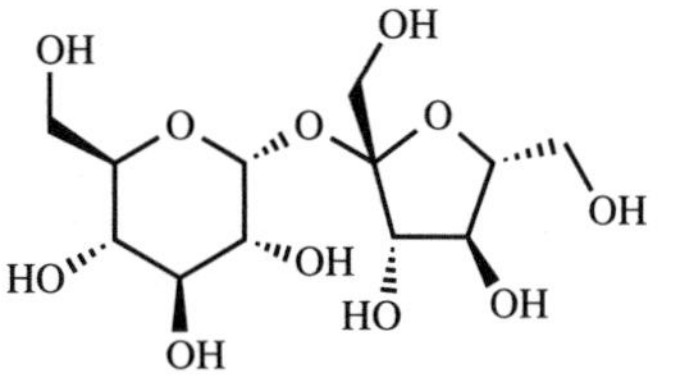

Sucrose
**1.23**

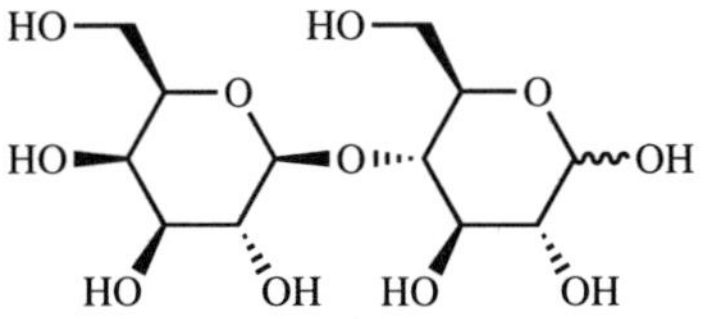

Lactose
**1.24**

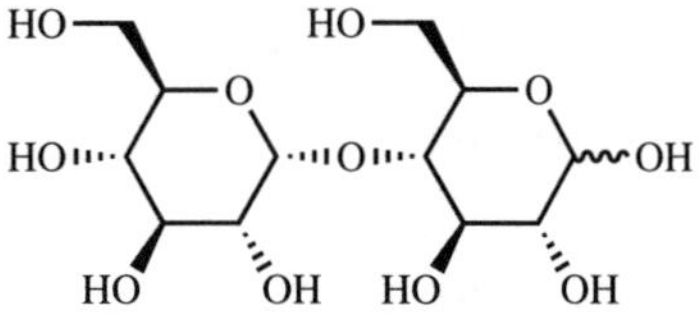

Maltose
**1.25**

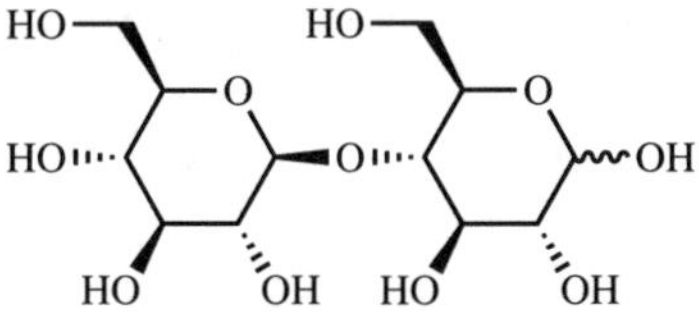

Cellobiose
**1.26**

## BUILDING BLOCK **2.01**

Name: α-D-Glucofuranose, 1,2:5,6-Bis-(1-methylethylidene)-[582-52-5]

**Synthesis:**

Acetone, $H^+$, 91%[1]

D-Glucose

**Related Structures:**

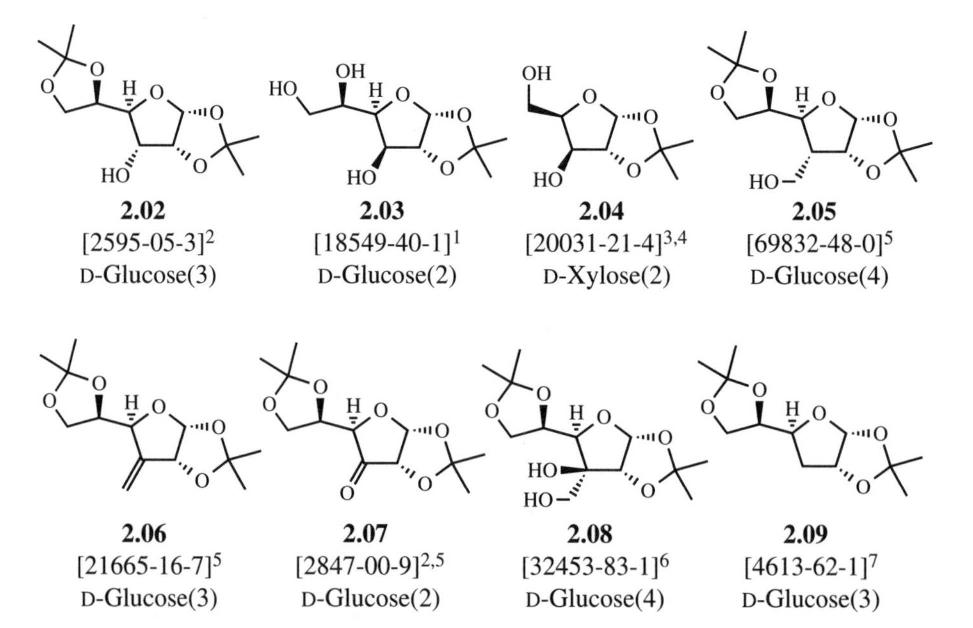

## BUILDING BLOCK **2.10**

Name: 1,3-Dioxolane-4-carboxaldehyde, 5-(Hydroxymethyl)-2,2-dimethyl (4*R*-*cis*)-[51607-16-0]

### Synthesis:

(a) BnOH, $H^+$, 91%[8]

(b) Acetone, $H^+$, 72%[8]

(c) $H_2$, Pd/C[8]

(d) $NaIO_4$, 77% (2 steps)[8]

D-Arabinose

### Related Structures:

**2.11** [6336-16-9][8] D-Arabinose(2)

**2.12** [18403-22-0][8] L-Arabinose(2)

**2.13** [4099-85-8][9] D-Ribose(1)

**2.14** [57492-90-7][10] D-Mannose(3)

**2.15** [4064-06-6][11] D-Galactose(1)

**2.16** [40269-01-0][12] D-Galactose(2)

**2.17** [14133-63-2][13] L-Rhamnose(2)

**2.18** [70266-87-4][14] D-Ribose(1)

## BUILDING BLOCK **2.19**

Name: $\alpha$-D-Galactopyranose, 4,6-*O*-Ethylidene-[13224-97-0]

### Synthesis:

D-Galactose → Paraldehyde, $H^+$, 58%[15]

### Related Structures:

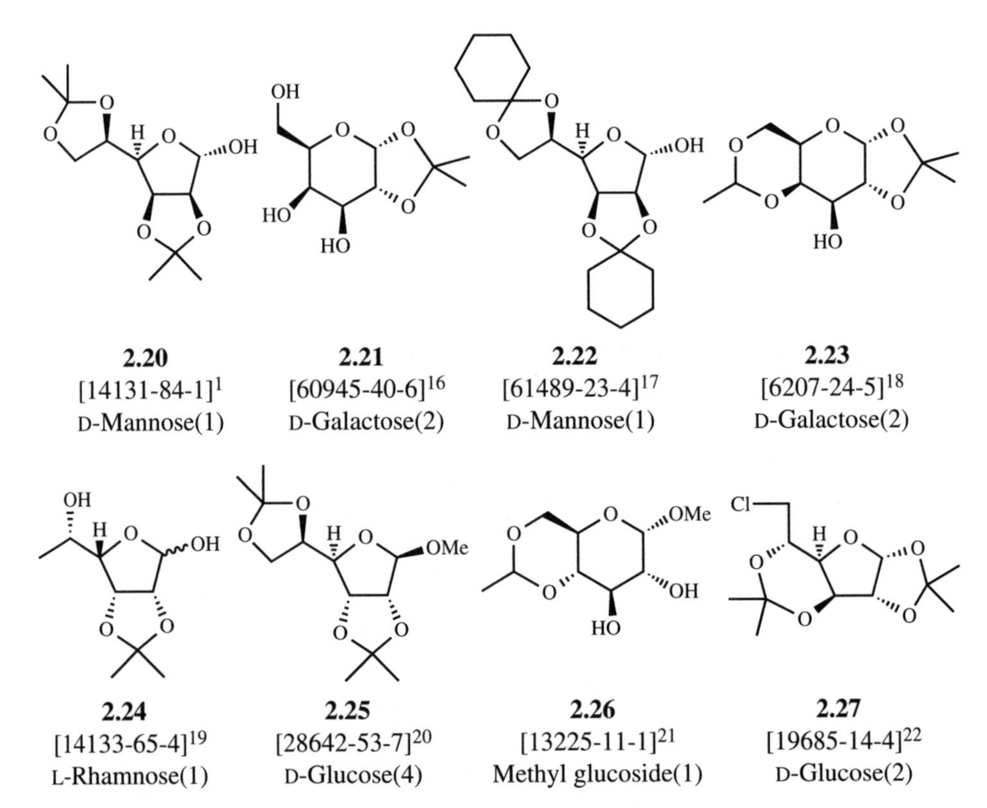

## BUILDING BLOCK **2.28**

Name: $\alpha$-D-Mannopyranose, 4,6-*O*-(1-Methylethylidene)-[68791-08-2]

### Synthesis:

D-Mannose

$CH_2{=}C(Me)OMe$, 82%[23]

TsOH, DMF, $CaCO_3$

### Related Structures:

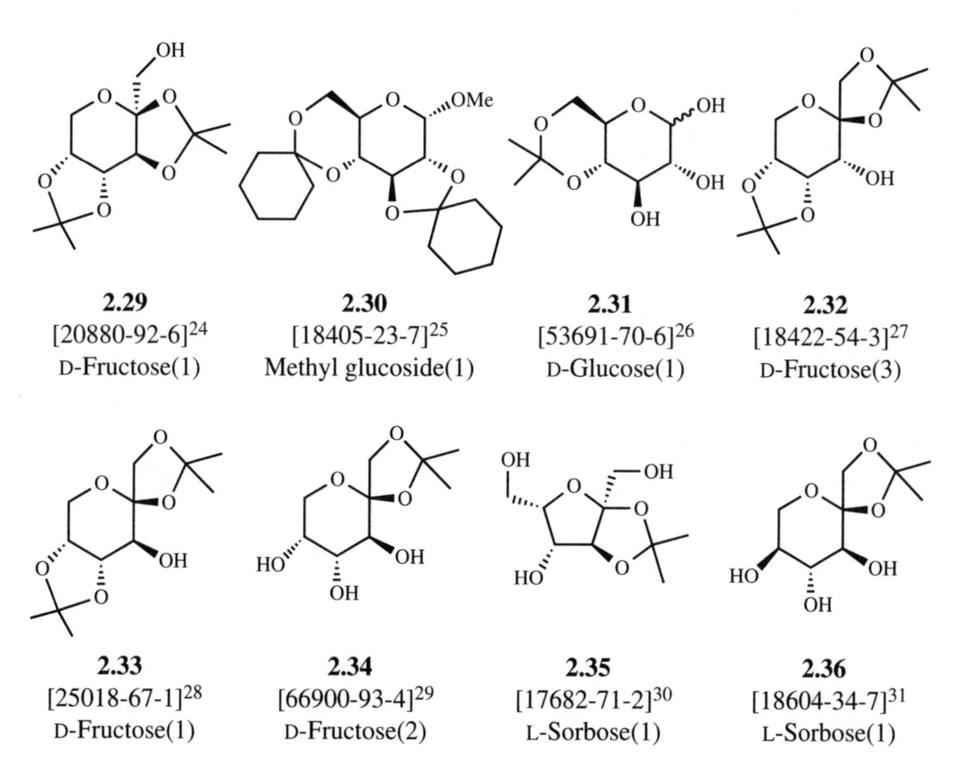

**2.29** [20880-92-6][24] D-Fructose(1)

**2.30** [18405-23-7][25] Methyl glucoside(1)

**2.31** [53691-70-6][26] D-Glucose(1)

**2.32** [18422-54-3][27] D-Fructose(3)

**2.33** [25018-67-1][28] D-Fructose(1)

**2.34** [66900-93-4][29] D-Fructose(2)

**2.35** [17682-71-2][30] L-Sorbose(1)

**2.36** [18604-34-7][31] L-Sorbose(1)

## BUILDING BLOCK **2.37**

Name: L-Idose, 2,5-Anhydro-dimethylacetal,
3,6-Bis(4-methylbenzenesulfonate)-[39022-34-9]

### Synthesis:

D-glucose

(a) Acetone, $H^+$, 91%[1]

(b) $H_3O^+$, 95%[1]

(c) TsCl, pyridine[32]

(d) MeOH, $H^+$, 95%[32]

### Related Structures:

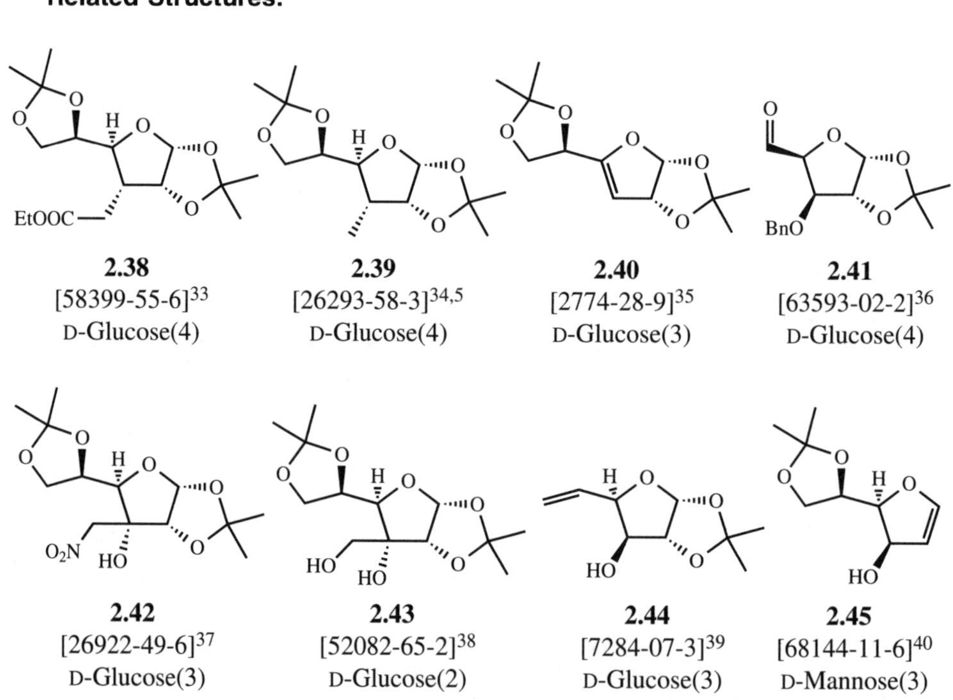

# BUILDING BLOCK **2.46**

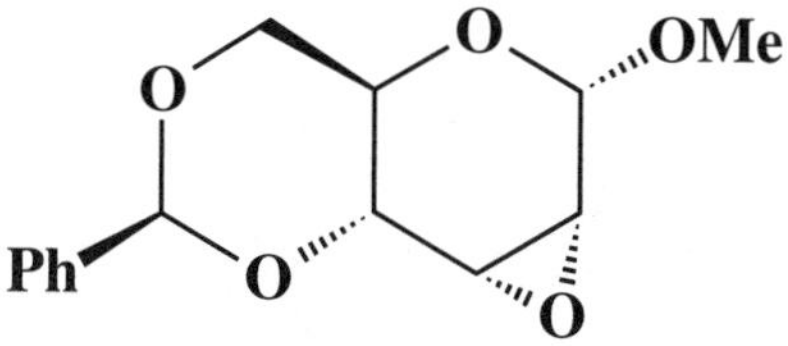

Name: α-D-Allopyranoside, Methyl 2,3-anhydro-4,6-(phenylmethylene)-[3150-15-0]

## Synthesis:

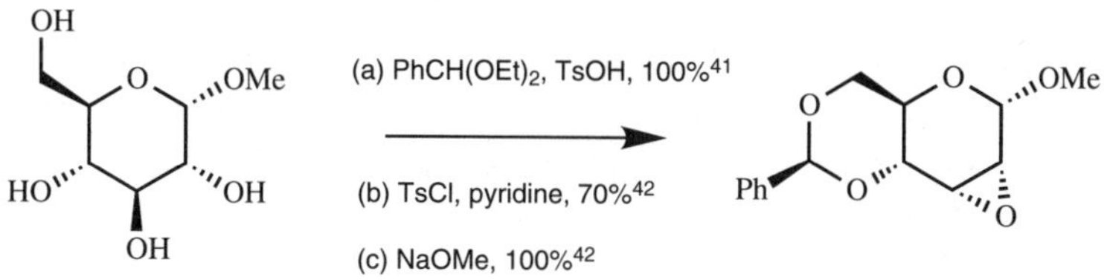

Methyl α-D-glucoside

(a) $PhCH(OEt)_2$, TsOH, 100%[41]

(b) TsCl, pyridine, 70%[42]

(c) NaOMe, 100%[42]

## Related Structures:

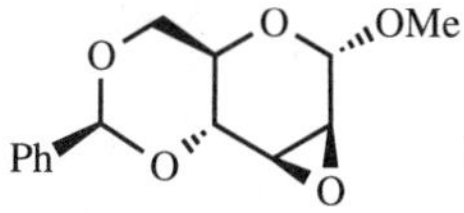

**2.47**
[3150-16-1][42]
Methyl glucoside(3)

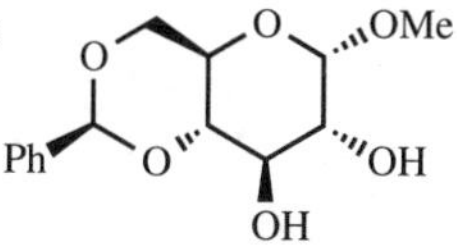

**2.48**
[3162-96-7][41]
Methyl glucoside(1)

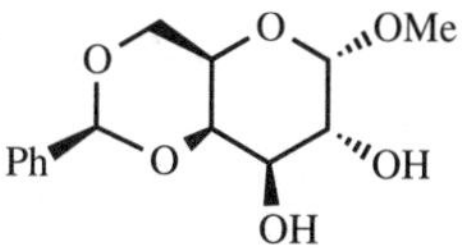

**2.49**
[4288-93-1][41]
D-Galactose(2)

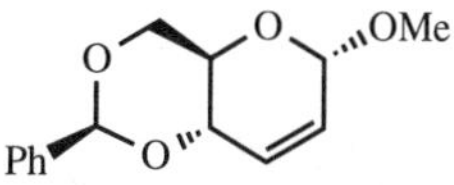

**2.50**
[3169-98-0][43]
Me glucoside(3)

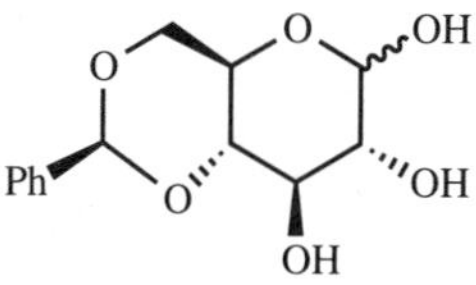

**2.51**
[25152-90-3][44]
D-Glucose(1)

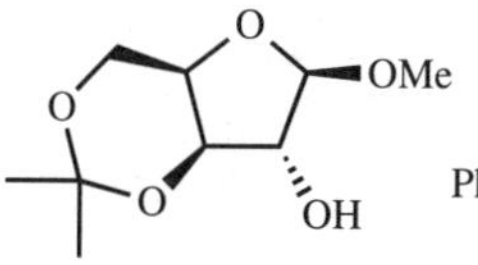

**2.52**
[51754-99-5][45]
D-Xylose(2)

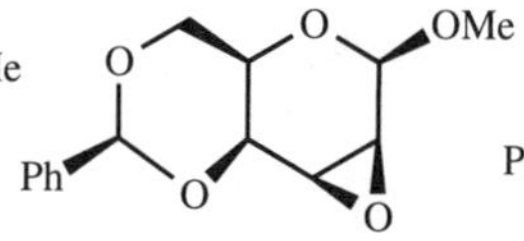

**2.53**
[15384-57-3][46]
D-Galactose(4)

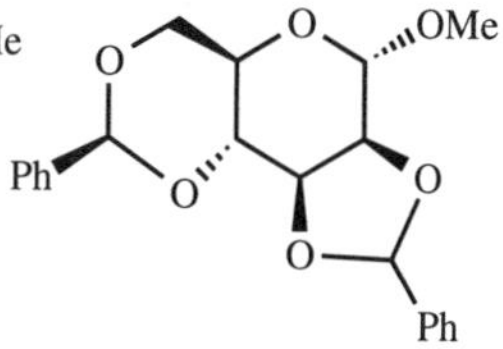

**2.54**
[4148-71-4][47]
D-Mannose(2)

## BUILDING BLOCK **2.55**

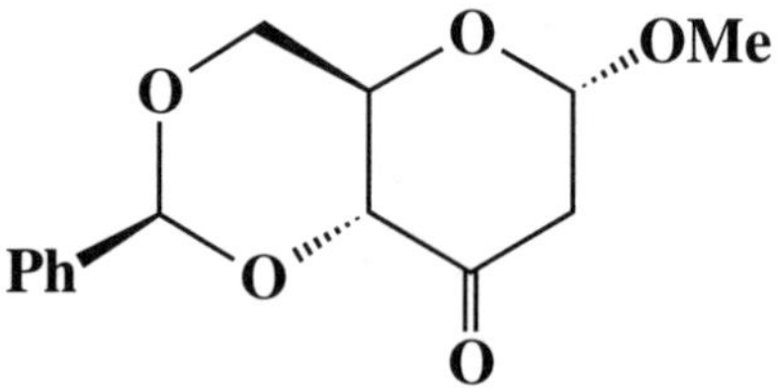

Name: α-D-*erythro*-Hexopyranosid-3-ulose, Methyl 2-deoxy-4,6-(phenylmethylene)-[6752-49-4]

### Synthesis:

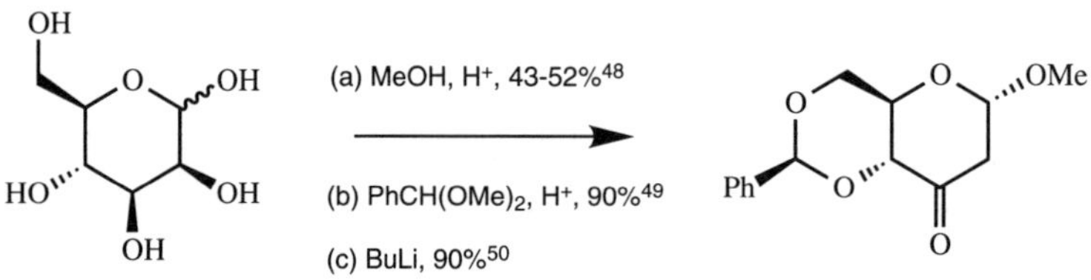

D-Mannose

(a) MeOH, $H^+$, 43-52%[48]

(b) $PhCH(OMe)_2$, $H^+$, 90%[49]

(c) BuLi, 90%[50]

### Related Structures:

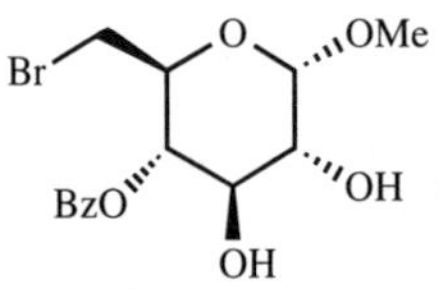

**2.56**
[10368-81-7][51]
Methyl glucoside(2)

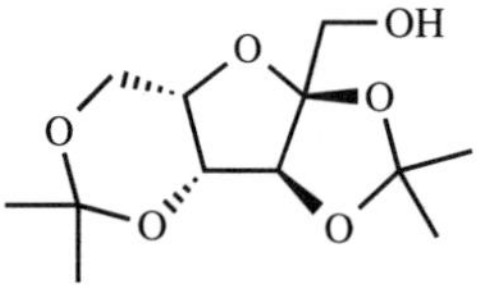

**2.57**
[17682-70-1][30]
L-Sorbose(1)

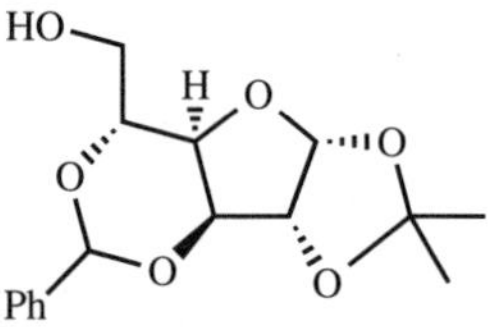

**2.58**
[22164-09-6][52]
D-Glucose(3)

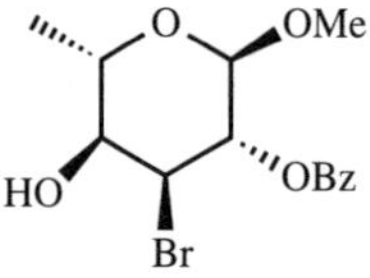

**2.59**
[64880-43-9][53]
L-Rhamnose(3)

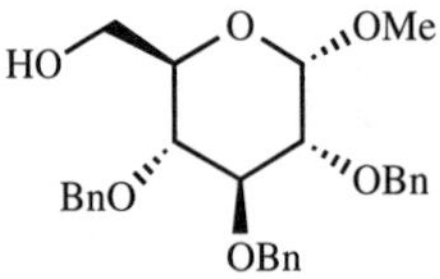

**2.60**
[53008-65-4][54]
Methyl glucoside(3)

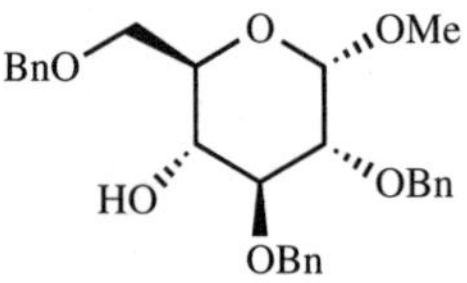

**2.61**
[19488-48-3][55]
Methyl glucoside(3)

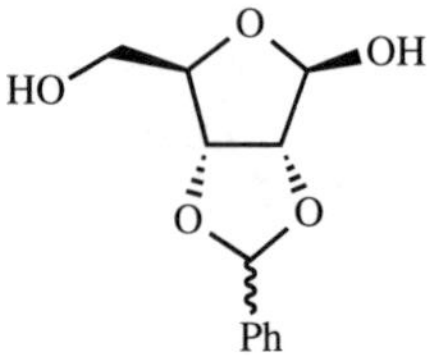

**2.62**
[39809-35-3][56]
D-Ribose(1)

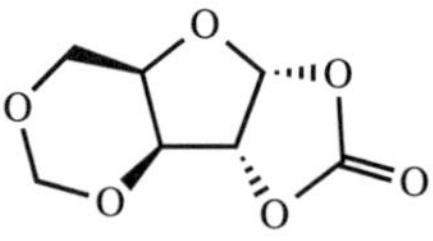

**2.63**
[—][57]
D-Xylose(2)

## BUILDING BLOCK **2.64**

OH H O OMe O O

Name: *β*-D-Allofuranoside, Methyl 6-deoxy-2,3-*O*-(1-methylethylidene)-[53270-44-3]

### Synthesis:

L-Rhamnose

(a) Acetone, $H^+$, 68%[19]

(b) TsCl, pyridine, 56%[19]

(c) NaOMe, 60%[19]

### Related Structures:

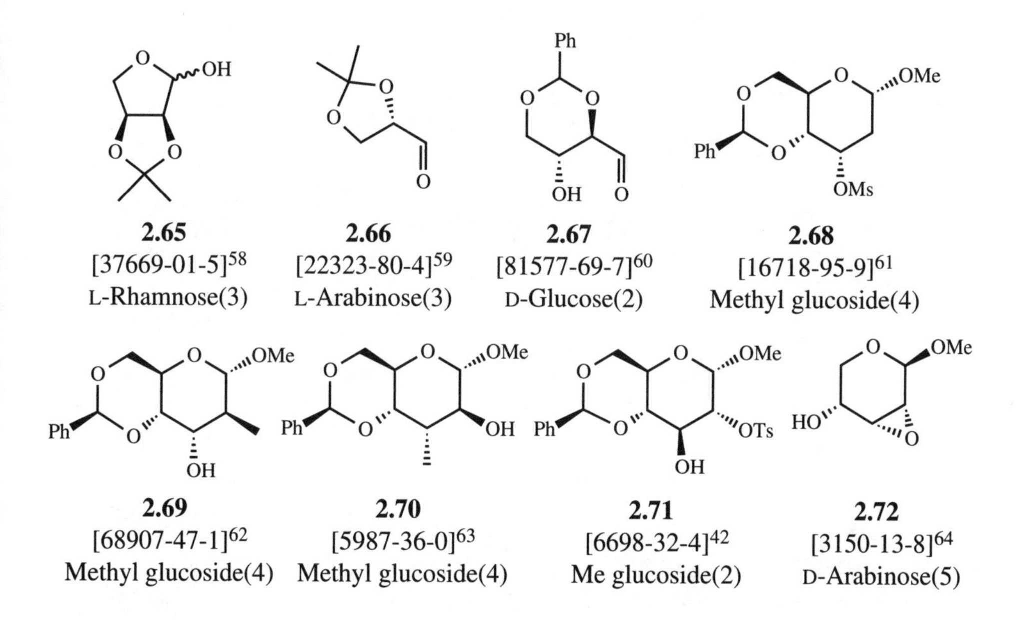

## BUILDING BLOCK **2.73**

HO OMe O O

Name: α-D-Lyxofuranoside, Methyl 2,3-anhydro-[26532-10-5]

### Synthesis:

D-Xylose

(a) MeOH, HCl[65]

(b) Acetone, $H^+$, 33-72% (2 steps)[65]

(c) MsCl, pyridine, 92%[65]

(d) AcOH, $H_2O$[65]

(e) NaOMe, 76% (2 steps)[65]

### Related Structures:

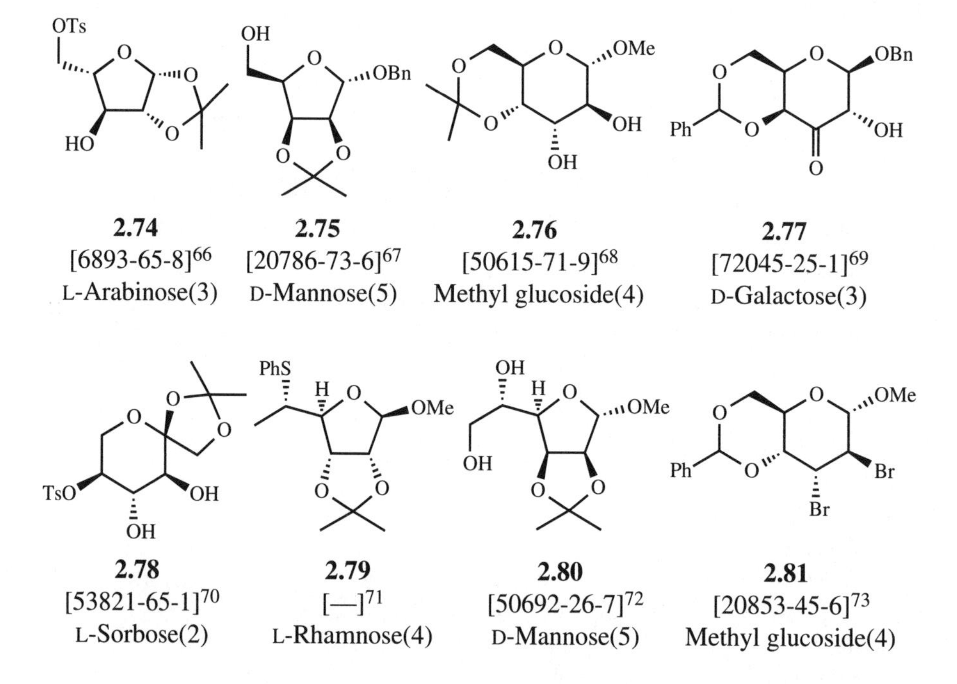

# BUILDING BLOCK **2.82**

Name: Hex-5-ulose-$\alpha$-D-*xylo*-furanuronic acid, 1,2-*O*-(1-Methylethylidene)-$\gamma$-lactone [5040-08-4]

## Synthesis:

(a) Acetone, $H^+$, 48%[74]

(b) $P_2O_5$, DMSO, 47%[75]

D-Glucuronolactone

## Related Structures:

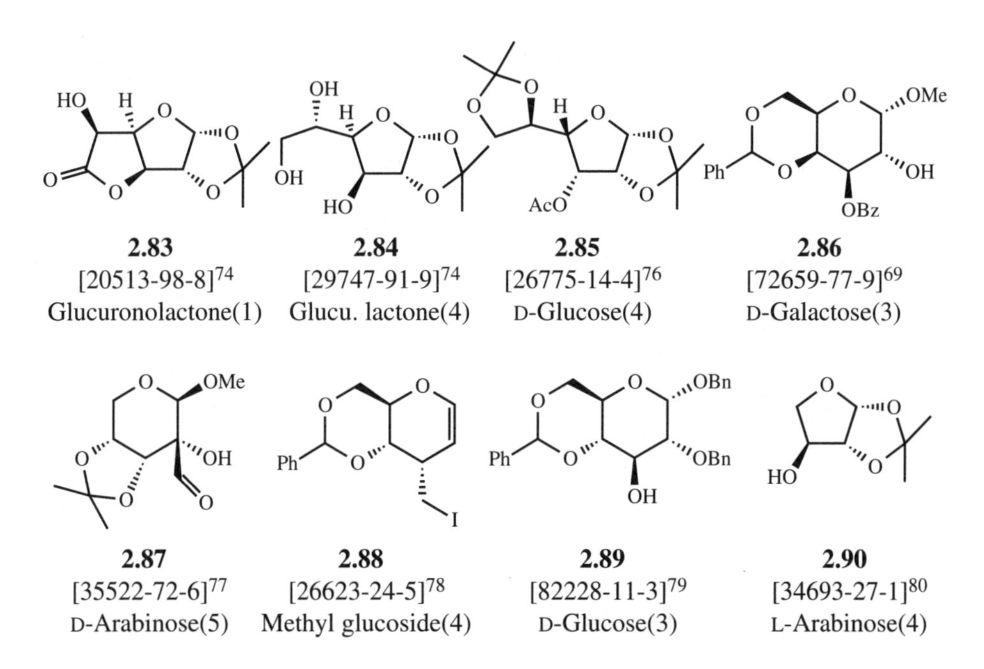

BUILDING BLOCK **2.91**

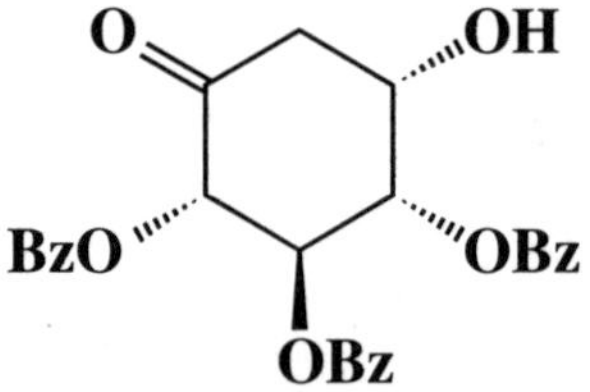

Name: Cyclohexanone, 2,3,4-Tris(benzoyloxy)-5-hydroxy-[2*S*-(2α,3β,4α,5α)]-[76371-39-6]

**Synthesis:**

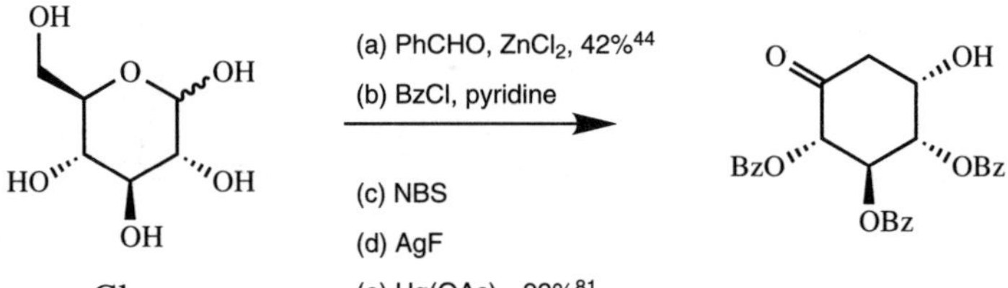

D-Glucose

(a) PhCHO, $ZnCl_2$, 42%[44]
(b) BzCl, pyridine
(c) NBS
(d) AgF
(e) $Hg(OAc)_2$, 93%[81]

**Related Structures:**

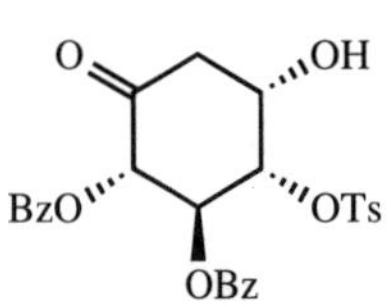

**2.92**
[72263-10-6][81]
Methyl glucoside(6)

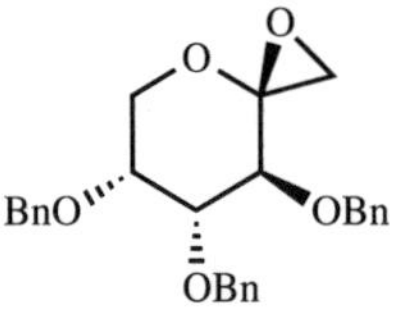

**2.93**
[154779-29-0][82]
D-Fructose(6)

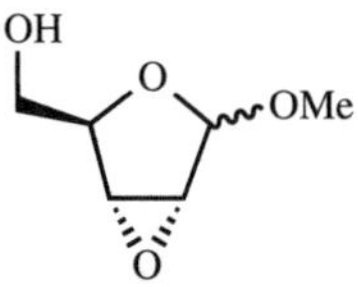

**2.94**
[4860-82-6][83]
D-Xylose(7)

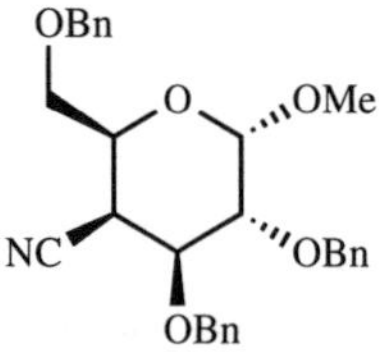

**2.95**
[—][84]
Me glucoside(5)

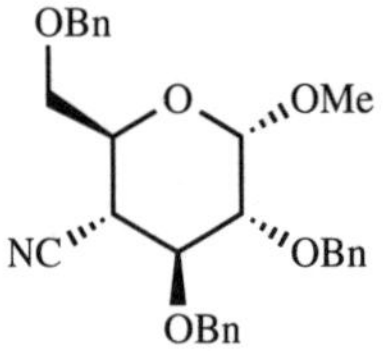

**2.96**
[—][84]
D-Galactose(6)

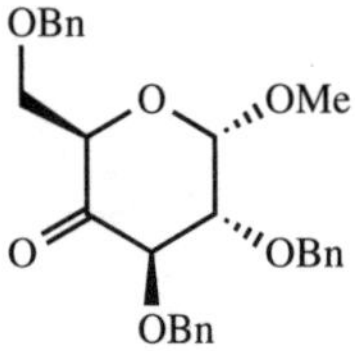

**2.97**
[98807-61-5][85]
Methyl glucoside(4)

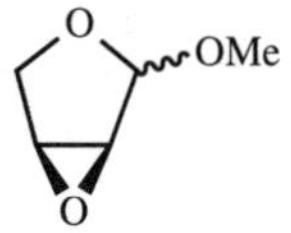

**2.98**
[148810-13-3][86]
D-Arabinose(7)

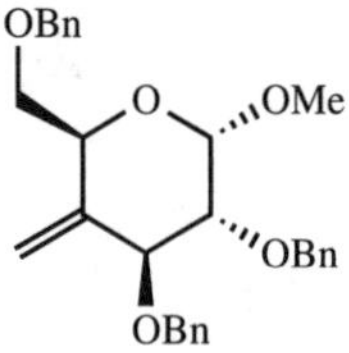

**2.99**
[139070-96-5][87]
Me glucoside(5)

## BUILDING BLOCK **2.100**

Name: D-*ribo*-Hex-1-entol, 1,5-Anhydro-2-deoxy-2-ethenyl-4,6-*O*-(phenylmethylene)-[117514-03-1]

### Synthesis:

Methyl glucoside

(a) $PhCH(OEt)_2$, $H^+$, 100%
(b) TsCl, pyridine, 70%
(c) NaOMe, 100%
(d) $LiCH(SMe)_2$, 75%
(e) $Ce(NH_4)_2(NO_3)_6$, 84%
(f) $Ph_3{=}CH_2$[88]

### Related Structures:

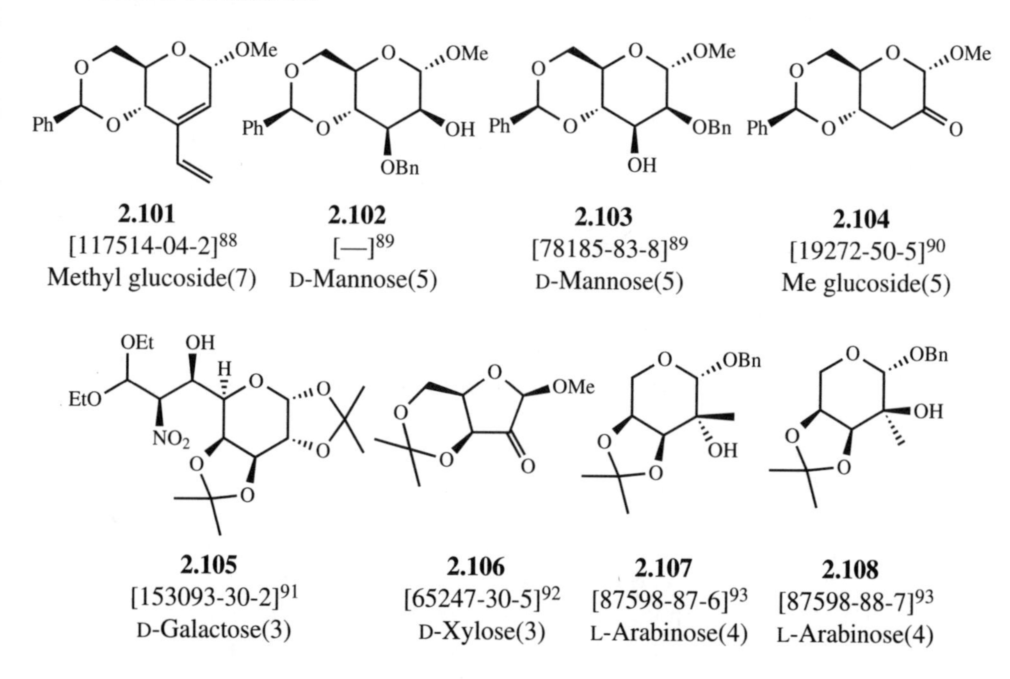

## REFERENCES

1. O.T. Schmidt. *Methods Carbohydr. Chem.* **1963** *2:*318–325.
2. J. D. Stevens. *Methods Carbohydr. Chem.* **1972** *6:*123–128.
3. P. A. Levene and A. L. Raymond. *J. Biol. Chem.* **1933** *102:*317–330.
4. O. Svanberg and K. Sjöberg. *Chem. Ber.* **1923** *56:*863–869.
5. A. Mazur, B. E. Tropp, and R. Engel. *Tetrahedron* **1984** *40:*3949–3956.
6. O. Dahlman, P. J. Garegg, H. Mayer, and S. Schramek. *Acta Chem. Scand. Ser. B* **1986** *40:*15–20.
7. D. H. R. Barton and S.W. McCombie. *J. Chem. Soc. Perkin I* **1975** 1574–1586.
8. C. E. Ballou. *J. Am. Chem. Soc.* **1957** *79:*165–166.
9. A. G. M. Barrett and S. A. Lebold. *J. Org. Chem.* **1990** *55:*3853–3857.
10. H. Ohrui and S. Emoto. *Tetrahedron* **1975** *32:*2765–2766.
11. R. S. Tipson. *Methods Carbohydr. Chem.* **1963** *2:*246–250.
12. M. L. Wolfrom, F. Shafizadeh, R. K. Armstrong, and T. M. Shen Hun. *J. Am. Chem. Soc.* **1959** *81:*3716–3719.
13. J. Jary, K. Capek, J. Kovar. *Collect. Czech. Chem. Commun.* **1963** *28:*2171–2181.
14. N. A. Hughes and P. R. H. Speakman. *Carbohydr. Res.* **1965** *1:*171–175.
15. D. H. Ball and J. K. N. Jones. *J. Chem. Soc.* **1958** 905–907.
16. A. M. Unrau. *Can. J. Chem.* **1963** *41:*2394–2397.
17. R. D. Guthrie and J. Honeyman. *J. Chem. Soc.* **1959** 853–854.
18. V. I. Betaneli, M.V. Ovchinnikov. L.V. Backinowsky, and N. K. Kochetkov. *Carbohydr. Res.* **1980** *84:*211–224.
19. B. R. Baker. *Methods Carbohydr. Chem.* **1963** *2:*441–444.
20. J. M. Williams. *Carbohydr. Res.* **1970** *13:*281–287.
21. D. M. Hall and O. A. Stamm. *Carbohydr. Res.* **1970** *12:*421–428.
22. S. Hanessian. *Methods Carbohydr. Chem.* **1972** *6:*190–193.
23. J. Gelas and D. Horton. *Carbohydr. Res.* **1978** *67:*371–387.
24. E. Pascu, E. J. Wilson, and L. Graf. *J. Am. Chem. Soc.* **1939** *61:*2675–2679.
25. F. H. Bisset, M. E. Evans, and F.W. Parish. *Carbohydr. Res.* **1967** *5:*184–193.
26. M. L. Wolfrom, A. B. Diwadkar, J. Gelas, and D. Horton. *Carbohydr. Res.* **1974** *35:*87–96.
27. S. M. Mio, Y. Kumagawa, and S. Sugai. *Tetrahedron* **1991** *47:*2133–2144.
28. K. Erne. *Acta Chem. Scand.* **1955** *9:*893–901.
29. W. A. Szarek, A. Zamojski, K. N. Tiwari, and E. R. Ison. *Tetrahedron Lett.* **1986** *27:*3827–3830.
30. T. Reichstein. *Helv. Chem. Acta* **1934** *17:*311–328.
31. H. Ohle. *Chem. Ber.* **1938** *71:*562–568.
32. T. Ogawa, M. Matsui, H. Ohrui, H. Kuzuhara, and S. Emoto. *Agr. Biol. Chem.* **1972** *36:*1449–1451.
33. G. J. Laurens and J. M. Koekemoer. *Tetrahedron Lett.* **1975** 3719–3722.
34. A. Rosenthal and M. Sprinzl. *Can. J. Chem.* **1969** *47:*3941–3946.
35. R. C. Anderson and B. Fraser-Reid. *Tetrahedron Lett.* **1977** 2865–2868.

36. M. L. Wolfrom and S. Hanessian. *J. Org. Chem.* **1962** *27:*1800–1804.
37. H. P. Albrecht and J. G. Moffatt. *Tetrahedron Lett.* **1970** 1063–1066.
38. W. P. Blackstock, C. C. Kuenzle, and C. H. Eugster. *Helv. Chem. Acta* **1974** *57:*1003–1009.
39. A.V. R. Rao, J. S. Yadav, C. S. Rao, and S. Chandraselchar. *J. Chem. Soc. Perkin I* **1990** 1211–1213.
40. R. E. Ireland, S. Thaisrivongs, N. Vanier, and C. S. Wilcox. *J. Org. Chem.* **1980** *45:*48–61.
41. V. Ferro, M. Mocerino, R.V. Stick, and D. M. G. Tilbrook. *Aust. J. Chem.* **1988** *41:*813–815.
42. L. F. Wiggins. *Methods Carbohydr. Chem.* **1963** *2:*188–191.
43. D. Horton, J. K. Thompson, and C. G. Tindall, Jr. *Methods Carbohydr. Chem.* **1972** *6:*297–301.
44. H. G. Fletcher. *Methods Carbohydr. Chem.* **1963** *2:*307–308.
45. E. E. Percival and R. Zobrist. *J. Chem. Soc.* **1952** 4306–4310.
46. L. F. Wiggins. *J. Chem. Soc.* **1944** 522–526.
47. D. Horton and W. Weckerle. *Carbohydr. Res.* **1975** *44:*227–240.
48. E. Fischer and L. Beensch. *Chem. Ber.* **1896** *29:*2927–2931.
49. N. L. Holder and B. Fraser-Reid. *Can. J. Chem.* **1973** *51:*3357–3365.
50. A. Klemer and G. Rodemeyer. *Chem. Ber.* **1974** *107:*2612–2614.
51. S. Hanessian. *Methods Carbohydr. Chem.* **1972** *6:*183–189.
52. J.T. Schmidt. *Methods Carbohydr. Chem.* **1962** *1:*199–200.
53. J. C. Florent, C. Monneret, and Q. Khuong-Huu. *Carbohydr. Res.* **1977** *56:*301–314.
54. K. Umemura, H. Matsuyama, M. Kobayshi, and N. Kamigata. *Bull. Chem. Soc. Japan* **1989** *62:*3026–3028.
55. P. J. Garegg and H. Hultberg. *Carbohydr. Res.* **1981** *93:*C10–C12.
56. T. B. Grindley and W. A. Szarek. *Carbohydr. Res.* **1972** *25:*187–195.
57. O.T. Schmidt. *Angew. Chem.* **1948** *60:*252.
58. J. N. Baxter and A. S. Perlin. *Can. J. Chem.* **1960** *38:*2217–2225.
59. S. B. Baker. *J. Am. Chem. Soc.* **1952** *74:*827–828.
60. S. R. Baker, D.W. Clissold, and A. McKillop. *Tetrahedron Lett.* **1988** *29:*991–994.
61. S. Kover, V. Dienstbierova, and J. Jary. *Collect. Czech. Chem. Commun.* **1967** *32:*2498–2503.
62. D. R. Hicks and B. Fraser-Reid. *Can. J. Chem.* **1975** *53:*2017–2023.
63. J.-R. Rougny and P. Sinay. *J. Chem. Res. (S)* **1982** *1:(M)* **1982** 0186–0196.
64. P.W. Kent, M. Stacey, and L. F. Wiggins. *J. Chem. Soc.* **1949** 1232–1235.
65. B. R. Baker, R. E. Schaub, and J. H. Williams. *J. Am. Chem. Soc.* **1955** *77:*7–12.
66. E. L. Hirst, J. K. N. Jones, and E. Williams. *J. Chem. Soc.* **1947** 1062–1064.
67. J. S. Brimacombe, F. Hunedy, and L. C. N. Tucker. *J. Chem. Soc. (C)* **1968** 1381–1384.
68. M. E. Evans. *Methods Carbohydr. Chem.* **1980** *8:*173–176.
69. S. David and A. Thieffy. *J. Chem. Soc. Perkin I* **1979** 1568–1573.
70. M. Chmielewski and R. L. Whistler. *J. Org. Chem.* **1975** *40:*639–643.

71. C. L. Stevens, R. P. Glinski, G. E. Gutowski, and J. P. Dickerson. *Tetrahedron Lett.* **1967** 649–652.

72. M. E. Evans and F.W. Parrish. *Methods Carbohydr. Chem.* **1980** *8:*173–176.

73. E. L. Albano, D. Horton, and J. H. Lauterbach. *Carbohydr. Res.* **1969** *9:*149–161.

74. M. Blanc-Meuesser, J. Defaye, D. Horton, and J.-H. Tsai. *Methods Carbohydr. Chem.* **1980** *8:*177–183.

75. K. Onodera and N. Kashimura. *Methods Carbohydr. Chem.* **1972** *6:*331–336.

76. K. N. Slessor and P. S. Tracey. *Can J. Chem.* **1969** *47:*3989–3995.

77. H. Paulsen, V. Sinnwell, and J. Thiem. *Methods Carbohydr. Chem.* **1980** *8:*185–194.

78. B. Fraser-Reid and B. Radatus. *Can. J. Chem.* **1969** *47:*4095–4097.

79. T. Ogawa and T. Kaburagi. *Carbohydr. Res.* **1982** *103:*53–64.

80. G. M. Blackburn and A. Rashid. *J. Chem. Soc. Chem. Commun.* **1988** 317–319.

81. R. Blattner, R. J. Ferrier, and S. R. Haines. *J. Chem. Soc. Perkin Trans. 1* **1985** 2413–2416.

82. M. M. Campbell and G. D. Heffernan. *Carbohydr. Res.* **1994** *251:*243–250.

83. C. D. Anderson, L. Goodman, and B. R. Baker. *J. Am. Chem. Soc.* **1958** *80:*5247–5252.

84. L. A. Reed III, J.T. Huang, M. McGregor, and L. Goodman. *Carbohydr. Res.* **1994** *254:*133–140.

85. S. M. Daly and R.W. Armstrong. *Tetrahedron Lett.* **1989** *30:*5713–5716.

86. J. Jary and I. Raich. *Carbohydr. Res.* **1993** *242:*291–295.

87. R. Preuss, K. H. Jung, and R. R. Schmidt. *Liebigs Ann. Chem.* **1992** 377–382.

88. J. C. Lopez, E. Lameignere, and G. Lukacs. *J. Chem. Soc. Chem. Commun.* **1988** 514–515.

89. Z. Szurmai, L. Balatoni, and A. Liptak. *Carbohydr. Res.* **1994** *254:*301–309.

90. A. Rosenthal and P. Catsoufacos. *Can. J. Chem.* **1969** *47:*2747–2750.

91. R. Fernandez, C. Gasch, A. Gomez-Sanchez, J.E. Vilchez, A. L. Castro, M. J. Dianez, M. D. Estrada, and S. Perez-Garrido. *Carbohydr. Res.* **1993** *247:*239–248.

92. K. Bischofberger, A. Brink, O. G. De Villiers, R. H. Hall, and A. Jordaan. *J. Chem. Soc. Perkin Trans. 1* **1977** 1472–1476.

93. R. E. Ireland, L. Courtney, and B. J. Fitzsimmons. *J. Org. Chem.* **1983** *48:*5186–5198.

## BUILDING BLOCK 3.01

Name: $\alpha$-D-Glucopyranoside, Methyl 2,6-dibenzoate [26927-44-6]

### Synthesis:

Methyl α-D-glucoside

$(Bu_3Sn)_2O$, BzCl, 95%[1]

### Related Structures:

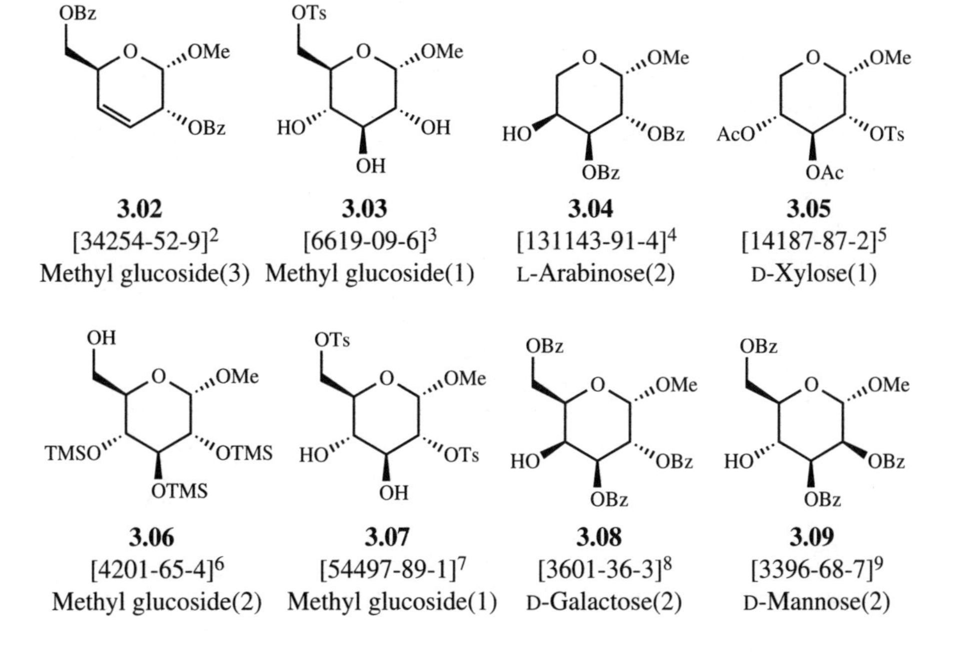

## BUILDING BLOCK **3.10**

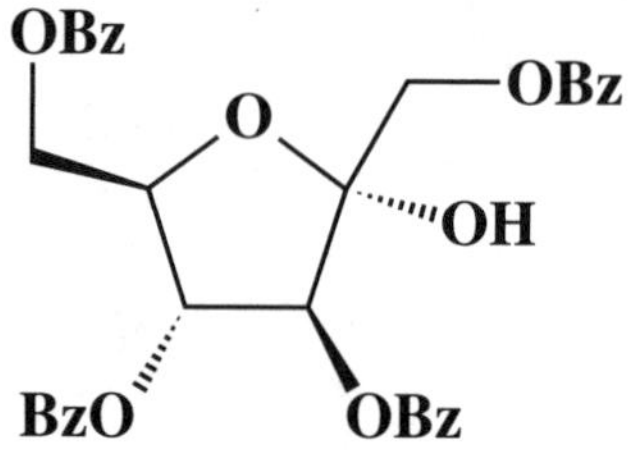
OBz
O
OBz
OH
BzO
OBz

Name: α-D-Fructofuranose, 1,3,4,6-Tetrabenzoate [80763-56-0]

### Synthesis:

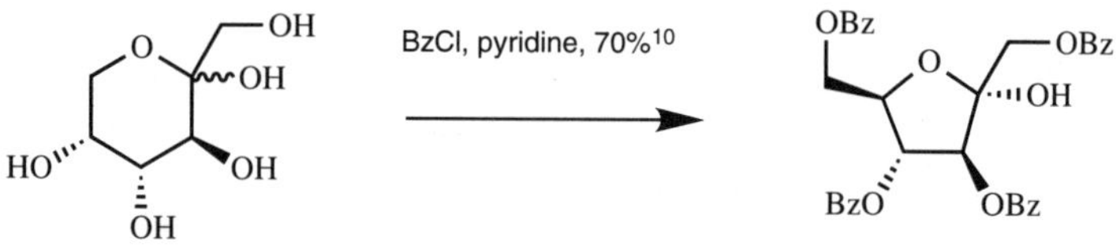
OH
O
OH
HO
OH
OH
BzCl, pyridine, 70%10
OBz
O
OBz
OH
BzO
OBz

D-Fructose

### Related Structures:

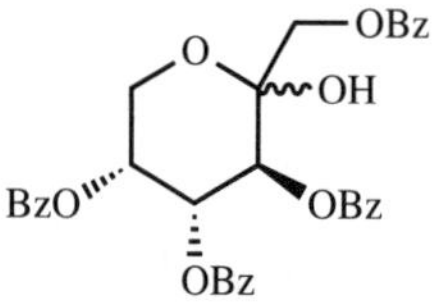
O
OBz
OH
BzO
OBz
OBz

**3.11**
[7143-89-7][11]
D-Fructose(1)

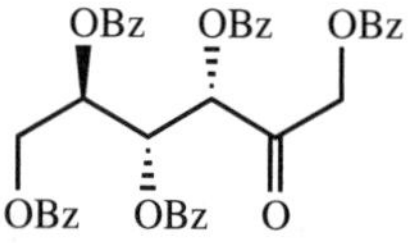
OBz
OBz
OBz
OBz
OBz
O

**3.12**
[83032-14-8][11]
D-Fructose(1)

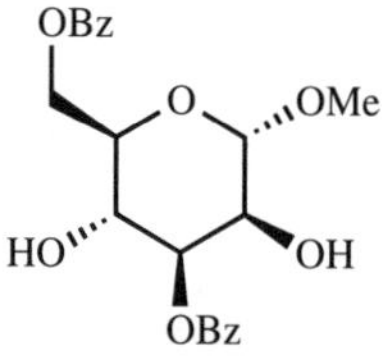
OBz
O
OMe
HO
OH
OBz

**3.13**
[14315-85-6][1]
D-Mannose(2)

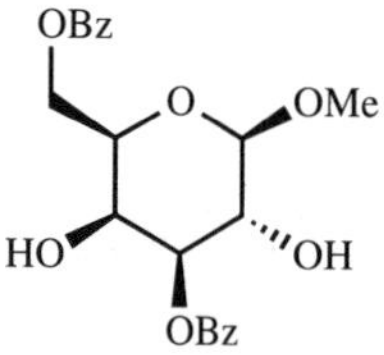
OBz
O
OMe
HO
OH
OBz

**3.14**
[64244-18-4][1]
D-Galactose(2)

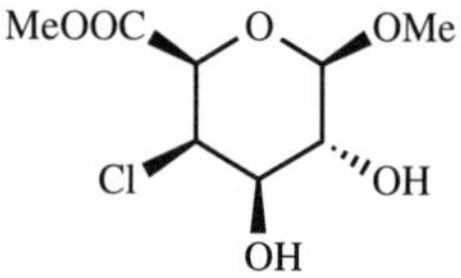
MeOOC
O
OMe
Cl
OH
OH

**3.15**
[—][12]
Glucuronolactone(2)

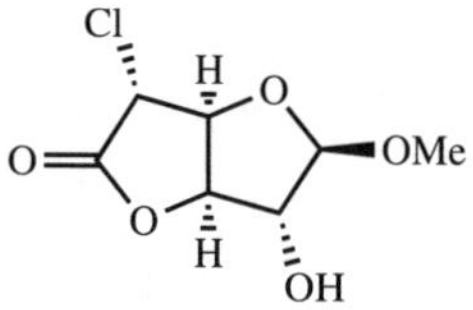
Cl
H
O
O
OMe
O
H
OH

**3.16**
[57569-49-0][13]
Glucuronolactone(1)

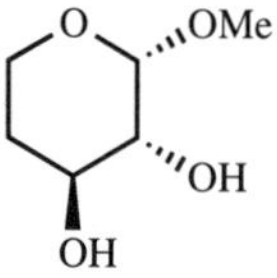
O
OMe
OH
OH

**3.17**
[23397-74-2][14]
D-Xylose(2)

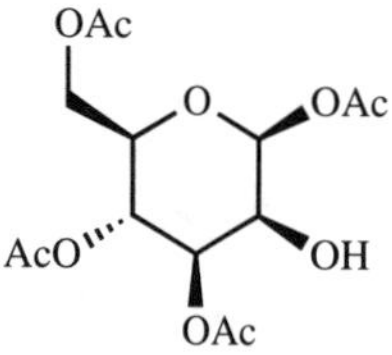
OAc
O
OAc
AcO
OH
OAc

**3.18**
[18968-05-3][15]
D-Mannose(1)

## BUILDING BLOCK **3.19**

Name: Piperidine, 1-(3,4,6-Tri-*O*-acetyl-*β*-D-glucopyranosyl)-[52389-39-6]

### Synthesis:

D-Glucose

(1) $Ac_2O$, NaOAc, 73%[16]

(2) piperidine, 36-62%[17]

### Related Structures:

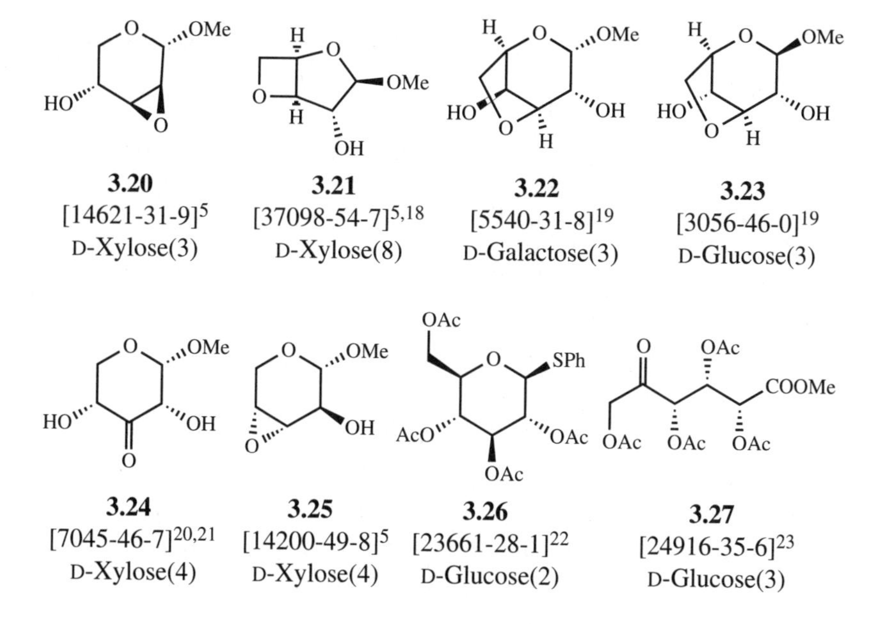

## BUILDING BLOCK 3.28

MeOOC O Br
AcO OAc
OAc

Name: α-D-Glucopyranuronic acid, 1-Bromo-1-deoxy-methyl ester, Triacetate [21085-72-3]

### Synthesis:

HO H O OH O O H OH

D-Glucuronolactone

(1) NaOMe[24]

(2) $Ac_2O$, 83% (2 steps)[24]

(3) HBr-AcOH, 85%[24]

MeOOC O Br AcO OAc OAc

### Related Structures:

OAc O Br AcO OAc OAc
OAc O Cl AcO OAc OAc
OAc O Cl AcO OH OAc
OAc O Cl AcO OH OAc

| **3.29** | **3.30** | **3.31** | **3.32** |
|---|---|---|---|
| [572-09-8][25] | [4451-36-9][26] | [4451-37-0][27] | [51295-68-2][28] |
| D-Glucose(1) | D-Glucose(2) | D-Glucose(3) | D-Galactose(3) |

OAc O O AcO OAc
OAc O O OMe O AcO OAc
OAc O OAc AcO OAc
O Cl MeO $OSO_2Cl$ $OSO_2Cl$

| **3.33** | **3.34** | **3.35** | **3.36** |
|---|---|---|---|
| [3867-86-5][27] | [4435-05-6][29] | [16750-06-4][30] | [—][31] |
| D-Glucose(3) | D-Mannose(2) | D-Glucose(2) | D-Xylose(3) |

## BUILDING BLOCK **3.37**

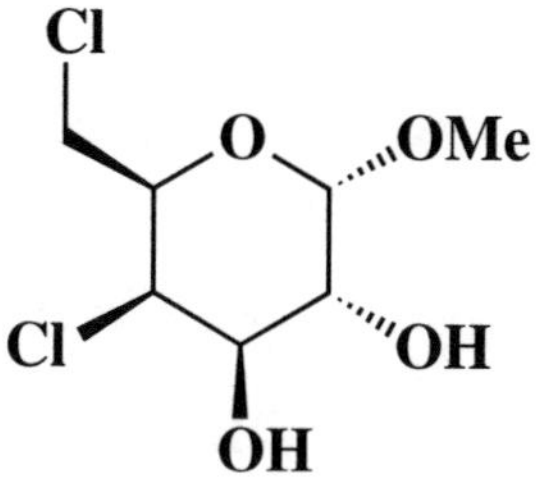

Name: α-D-Galactopyranoside, Methyl 4,6-dichloro-4,6-dideoxy-[4990-82-3]

### Synthesis:

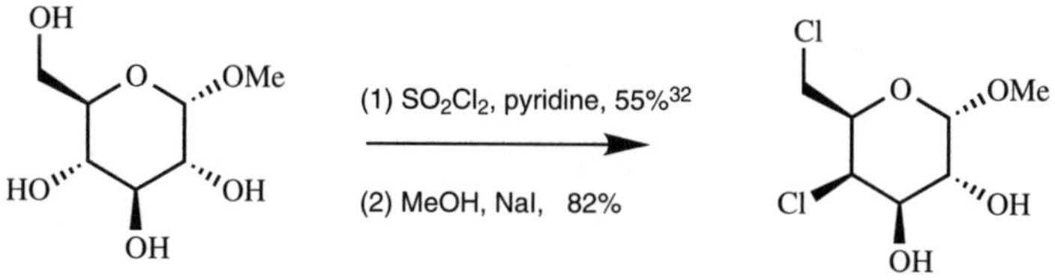

Methyl glucoside

(1) $SO_2Cl_2$, pyridine, 55%[32]

(2) MeOH, NaI, 82%

### Related Structures:

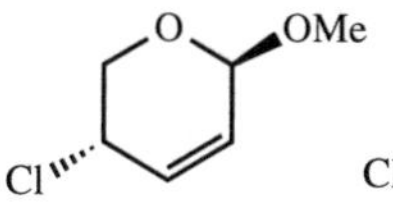

**3.38** [—][33] L-Arabinose(3)

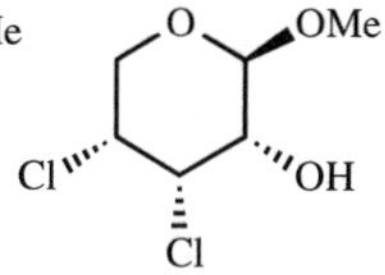

**3.39** [—][33] L-Arabinose(3)

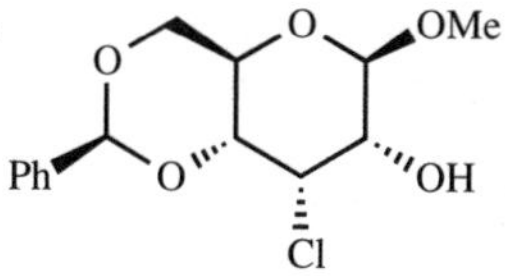

**3.40** [4990-99-2][34] D-Glucose(3)

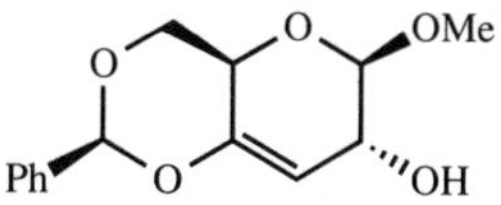

**3.41** [34340-38-0][34] D-Glucose(4)

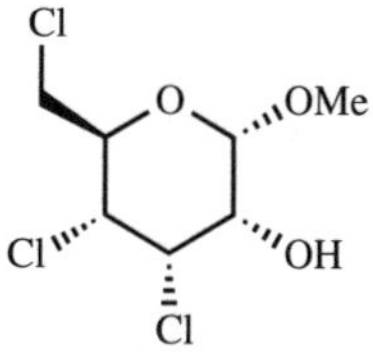

**3.42** [—][35] D-Galactose(2)

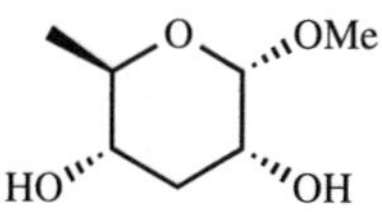

**3.43** [31899-66-8][36] Methyl glucoside(2)

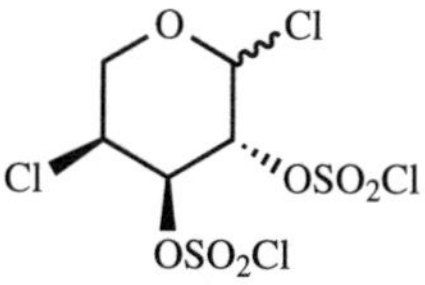

**3.44** [—][37] D-Xylose(1)

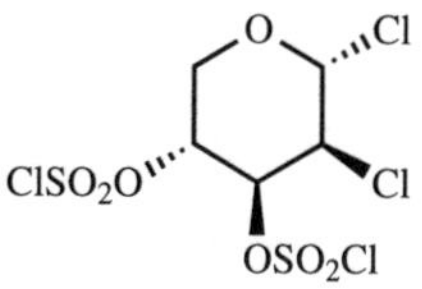

**3.45** [29217-59-2][38] D-Xylose(2)

## BUILDING BLOCK **3.46**

Name: D-Gluconic acid, 2-*C*-Bromo-$\delta$-lactone tetrabenzoate [76514-09-5]

### Synthesis:

D-Glucose

(1) BzCl, base, 57%
(2) HBr-HOAc, 94%
(3) MeOH, $Ag_2O$, 69%
(4) NBS, 48 %[39]

### Related Structures:

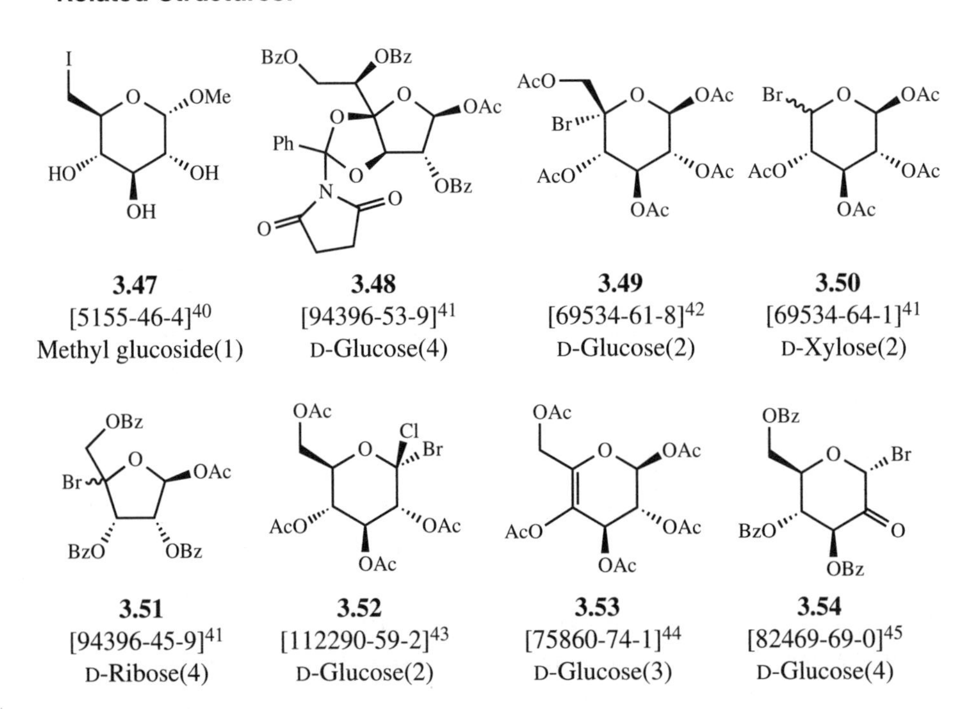

**3.47** [5155-46-4][40] Methyl glucoside(1)

**3.48** [94396-53-9][41] D-Glucose(4)

**3.49** [69534-61-8][42] D-Glucose(2)

**3.50** [69534-64-1][41] D-Xylose(2)

**3.51** [94396-45-9][41] D-Ribose(4)

**3.52** [112290-59-2][43] D-Glucose(2)

**3.53** [75860-74-1][44] D-Glucose(3)

**3.54** [82469-69-0][45] D-Glucose(4)

## BUILDING BLOCK **3.55**

OH

O OBn

O OBn

OBn

Name: $\alpha$-D-*xylo*-Hexopyranosid-4-ulose, Phenylmethyl 2,3-bis-*O*-(phenylmethyl)-[72045-22-8]

### Synthesis:

D-Glucose

(1) BnOH, H+ 60%
(2) PhCHO, H+, 50%
(3) BnCl, NaH, 91%
(4) $H_3O^+$, 80-85%
(5) $Bu_2SnO$, $Br_2$, 87%[46]

### Related Structures:

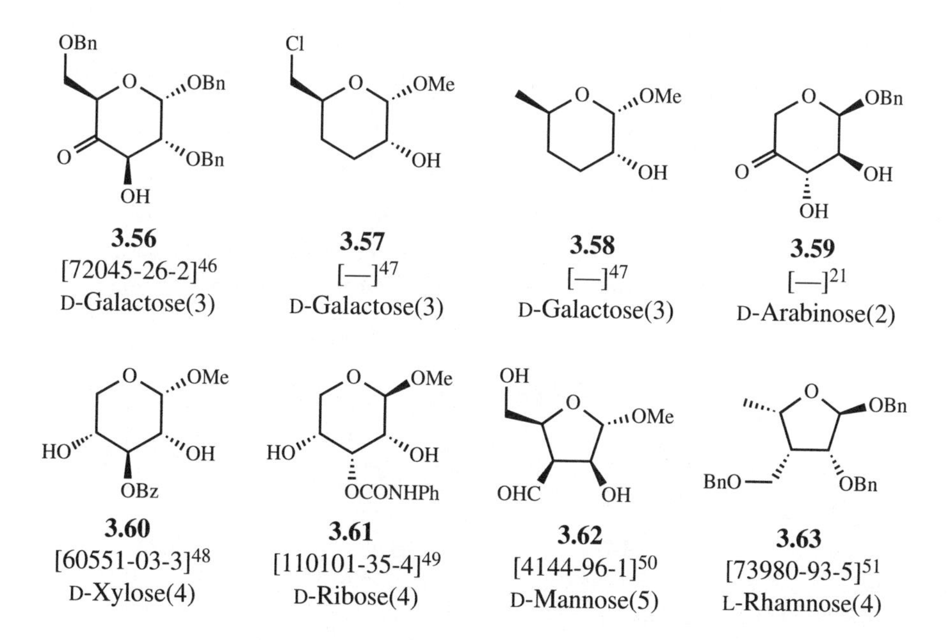

## BUILDING BLOCK **3.64**

OBn
OMe
BnO
OBn
OH

Name: $\alpha$-D-Glucopyranoside, Methyl 2,4,6-Tris-*O*-(phenylmethyl)-[35303-86-7]

### Synthesis:

BnCl[52]
KOH
100°C
3 hr, 50%

Methyl glucoside

### Related Structures:

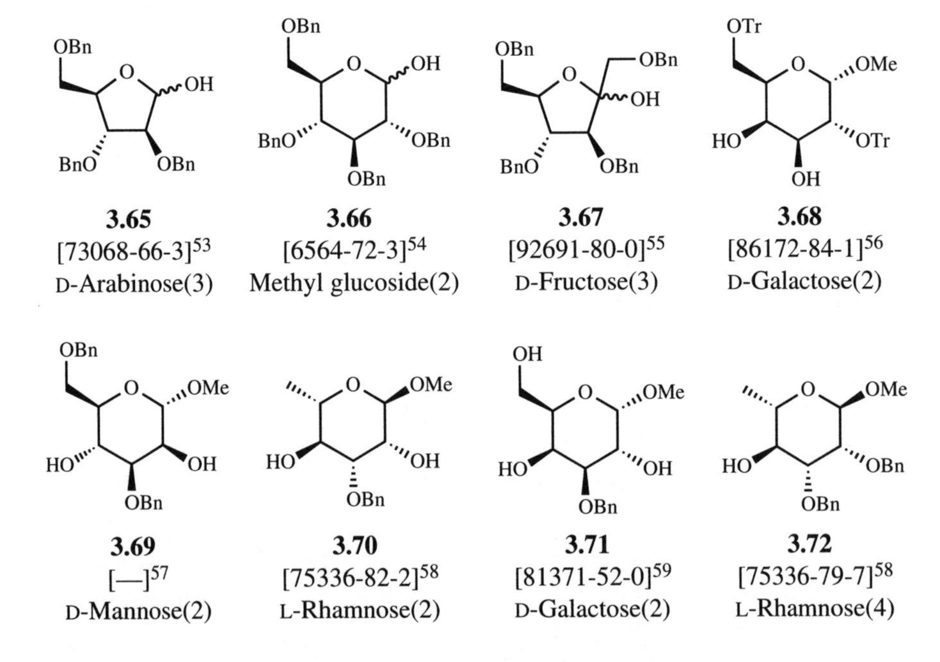

## BUILDING BLOCK **3.73**

BnO OBn O O

Name: $\alpha$-D-Glucopyranose, 1,2-Anhydro-6-deoxy-3,4-bis-*O*-(phenylmethyl)-
[154779-60-9]

### Synthesis:

L-Rhamnose

(a) $Ac_2O$, $Br_2$, P
(b) MeCN, $NaBH_4$, 97%
(c) BnBr, KOH
(d) $H_3O^+$
(e) TsCl, $K_2CO_3$, 61%
(f) $KOBu^t$, 94%[60]

### Related Structures:

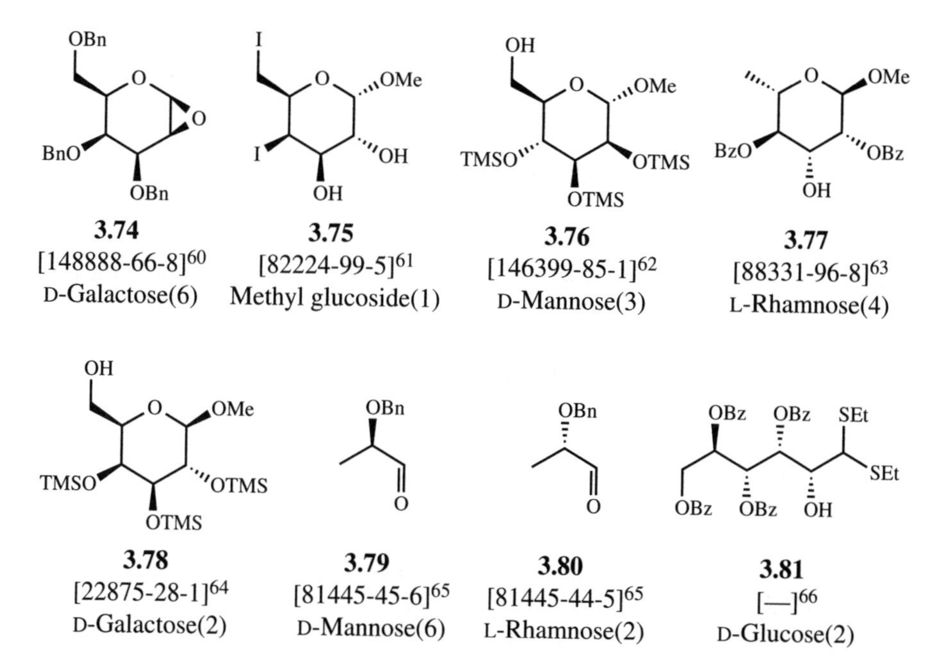

## REFERENCES


1. T. Ogawa and M. Matsui. *Tetrahedron* **1981** *37:*2363–2369.
2. N. L. Holder and B. Fraser-Reid. *Can. J. Chem.* **1973** *51:*3357–3365.
3. F. Cramer, H. Otterbach, and H. Springbaum. *Chem. Ber.* **1959** *92:*384–391.
4. E. J. Reist, L.V. Fisher, and L. Goodman. *J. Org. Chem.* **1967** *32:*2541–2544.
5. J. G. Buchanan. *Methods Carbohydr. Chem.* **1972** *6:*135–141.
6. D.T. Hurst and A. G. McInnes. *Can. J. Chem.* **1965** *43:*2004–2011.
7. J. Jary, K. Capek, and J. Kovar. *Collect. Czech. Chem. Commun.* **1964** *29:*930–937.
8. E. J. Reist, R. R. Spencer, D. F. Calkins, B. R. Baker, and L. Goodman. *J. Org. Chem.* **1965** *30:*2312–2317.
9. A. C. Richardson and J. M. Williams. *Tetrahedron* **1967** *23:*1641–1646.
10. P. Briegl and R. Schinle. *Chem. Ber.* **1934** *67:*127–130.
11. P. Briegl and R. Schinle. *Chem. Ber.* **1933** *66:*325–330.
12. S. S. Thomas, J. Plenkiewiz, E. R. Ison, M. Bols, W. Zou, W. A. Szarek, and R. Kisilevsky. *Biochem. Biophys. Acta* **1995** (*in press*).
13. H. Parolis. *Carbohydr. Res.* **1983** *114:*21–33.
14. K. E. Dow, R. J. Riopelle, W. A. Szarek, M. Bols, E. R. Ison, J. Plenkiewicz, A. Lyon, and R. Kisilevsky. *Biochem. Biophys. Acta* **1992** *1156:*7–14.
15. J. O. Deferrari, E. G. Gros, and I. M. E. Thiel. *Methods Carbohydr. Chem.* **1972** *6:*365–367.
16. M. L. Wolfrom and A. Thompson. *Methods Carbohydr. Chem.* **1963** *2:*211–215.
17. J. E. Hodge and C. E. Rist. *J. Am. Chem. Soc.* **1952** *74:*1498–1500.
18. C. D. Anderson, L. Goodman, and B. R. Baker. *J. Am. Chem. Soc.* **1958** *80:*5247–5252.
19. B. A. Lewis, F. Smith, and A. M. Stephen. *Methods Carbohydr. Chem.* **1963** *2:*172–187.
20. B. Lindberg. *Methods Carbohydr. Chem.* **1972** *6:*323–325.
21. K. Heyns and J. Lenz. *Angew. Chem.* **1961** *73:*299.
22. R. J. Ferrier, R. H. Furneaux. *Methods Carbohydr. Chem.* **1980** *8:*251–253.
23. S. J. Angyal and K. James. Australian patent 417,759, 4 Nov. **1971**.
24. G. N. Bollenback, J.W. Long, D. G. Benjamin, and J. A. Lindquist. *J. Am. Chem. Soc.* **1955** *77:*3310–3315.
25. R. U. Lemieux. *Methods Carbohydr. Chem.* **1963** *2:*221–222.
26. K. Heyns, W. P. Trautwein, F. G. Espinoza, and H. Paulsen. *Chem. Ber.* **1966** *99:*1183–1191.
27. R. U. Lemieux and J. Howard. *Methods Carbohydr. Chem.* **1963** *2:*400–402.
28. P. M. Collins, W. G. Overend, and B. A. Rayner. *Carbohydr. Res.* **1973** *31:*1–16.
29. H. B. Boren, G. Ekborg, K. Eklind, P. J. Garegg, A. Pilotti, and C. G. Swahn. *Acta Chem. Scand.* **1973** *27:*2639–2644.
30. B. Giese, S. Gilges, K. S. Groeninger, C. Lamberth, and T. Witzel. *Liebigs. Ann. Chim.* **1988** 615–617.
31. C. F. Gibbs and H. J. Jennings. *Can. J. Chem.* **1970** *48:*2735–2739.

32. H. J. Jennings and J. K. N. Jones. *Can. J. Chem.* **1963** *41:*1151–1159.
33. H. J. Jennings and J. K. N. Jones. *Can. J. Chem.* **1965** *43:*3018–3025.
34. E. H. Williams, W. A. Szarek, and J. K. N. Jones. *Carbohydr. Res.* **1971** *20:*49–57.
35. H. J. Jennings and J. K. N. Jones. *Can. J. Chem.* **1965** *43:*2372–2386.
36. G. Ekborg and S. Svenson. *Acta Chem. Scand.* **1973** *27:*1437–1439.
37. H. J. Jennings and J. K. N. Jones. *Can. J. Chem.* **1962** *40:*1408–1414.
38. H. J. Jennings. *Can. J. Chem.* **1970** *48:*1834–1841.
39. R. J. Ferrier and P. C. Tyler. *J. Chem. Soc. Perkin I* **1980** 2762–2766.
40. R. L. Whistler and A. K. M. Anisuzzaman. *Methods Carbohydr. Chem.* **1980** *8:*227–231.
41. R. J. Ferrier, S. R. Haines, G. J. Gainsford, and E. J. Gabe. *J. Chem. Soc. Perkin I* **1984** 1683–1687.
42. R. Blattner and R. J. Ferrier. *J. Chem. Soc. Perkin I* **1980** 1523–1527.
43. J.-P. Praly, L. Brard, G. Descotes, and L. Toupet. *Tetrahedron* **1989** *45:*4141–4152.
44. R. Blattner, R. J. Ferrier, and P. C. Tyler. *J. Chem. Soc. Perkin I* **1980** 1535–1539.
45. F.W. Lichtenthaler, P. Jarglis, and W. Hempe. *J. Liebigs. Ann. Chim.* **1983** 1959–1972.
46. S. David and A. Thieffry. *J. Chem. Soc. Perkin I* **1979** 1568–1573.
47. B.T. Lawton, W. A. Szarek, and J. K. N. Jones. (*unpublished*); W. A. Szarek. *Adv. Carbohydr. Chem. Biochem.* **1973** *28:*225.
48. R. J. Ferrier, D. Prasard, A. Rudowski, and I. Sangster. *J. Chem. Soc.* **1964** 3330–3334.
49. R. J. Ferrier and D. Prasad. *J. Chem. Soc.* **1965** 7425–7428.
50. P.W. Austin, J. G. Buchanan, and R. M. Saunders. *J. Chem. Soc. (C)* **1967** 372–377.
51. V. Pozsgay and A. Neszmelyi. *Tetrahedron Lett.* **1980** *21:*211–212.
52. S. Koto, N. Morishima, M. Owa, and S. Zen. *Carbohydr. Res.* **1984** *130:*73–83.
53. S. Tejima and H. G. Fletcher, Jr. *J. Org. Chem.* **1963** *28:*2999–3004.
54. C. P. J. Glaudemans and H. G. Fletcher, Jr. *Methods Carbohydr. Chem.* **1972** *6:*373–376.
55. R. K. Ness, H.W. Diehl, and H. G. Fletcher, Jr. *Carbohydr. Res.* **1970** *13:*23–32.
56. T. Ogawa, T. Nukada, and M. Matsui. *Carbohydr. Res.* **1982** *101:*263–270.
57. T. Ogawa and M. Matsui. *Carbohydr. Res.* **1978** *62:*C1-C4.
58. S. S. Rana, J. J. Barlow, K. L. Matta. *Carbohydr. Res.* **1980** *85:*313–317.
59. R. R. Conteras, J. P. Kamerling, J. Bres, and J. F. G. Vliegenthart. *Carbohydr. Res.* **1988** *179:*411–418.
60. E. Wu and Q. Wu. *Carbohydr. Res.* **1993** *250:*327–333.
61. P. J. Garregg, R. Johansson, C. Ortega, and B. Samuelsson. *J. Chem. Soc. Perkin Trans. 1* **1982** 681–683.
62. M. Meldal, M. K. Christensen, and K. Bock. *Carbohydr. Res.* **1992** *235:*115–127.
63. H. P. Wessel and D. R. Bundle. *Carbohydr. Res.* **1983** *124:*301–311.
64. R. F. Helm, J. Ralph, and R. D. Hatfield. *Carbohydr. Res.* **1992** *229:*183–194.
65. D. C. Baker and L. D. Hawkins. *J. Org. Chem.* **1982** *47:*2172–2184.
66. H. R. Bolliger and D. M. Schmid. *Methods Carbohydr. Chem.* **1962** *1:*186–189.

## BUILDING BLOCK **4.01**

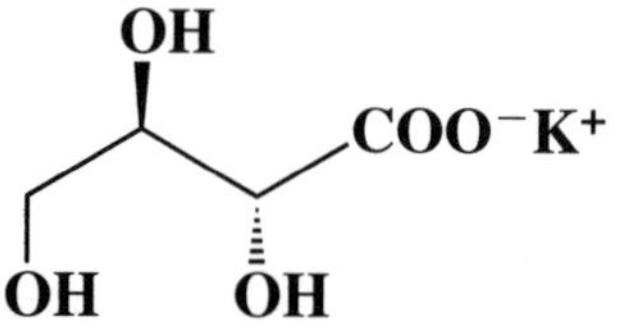
OH
COO⁻K⁺
OH OH

Name: Butanoic acid, 2,3,4-Trihydroxy-monopotassium salt (3*R*,4*R*)-
[88759-55-1]

### Synthesis:

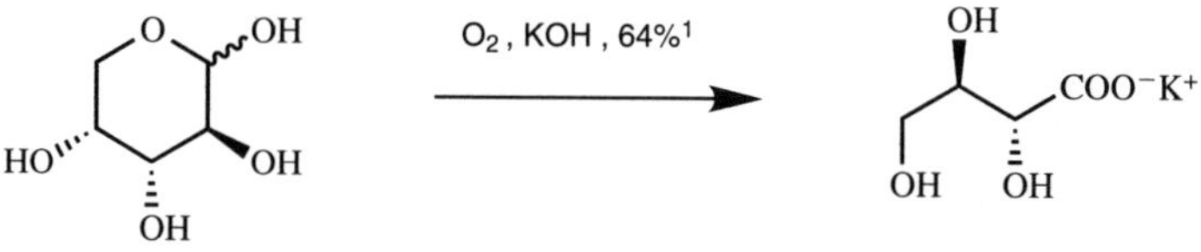
O
OH
HO
OH
OH
O2, KOH, 64%1
OH
COO⁻K⁺
OH OH

D-Arabinose

### Related Structures:

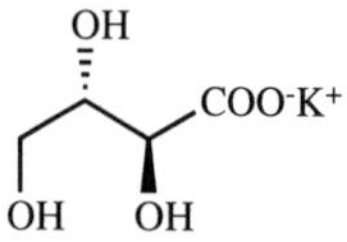

**4.02**
[88759-57-3][1]
L-Arabinose(1)

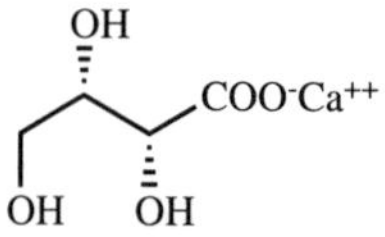

**4.03**
[70753-61-6][2]
L-Ascorbic acid(1)

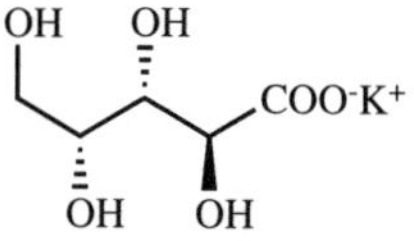

**4.04**
[78138-87-1][1]
D-Galactose(1)

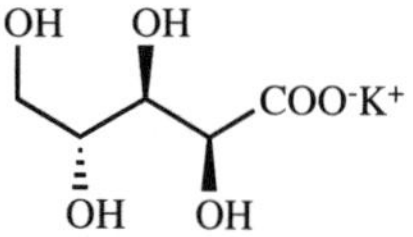

**4.05**
[15770-22-6][1]
D-Glucose(1)

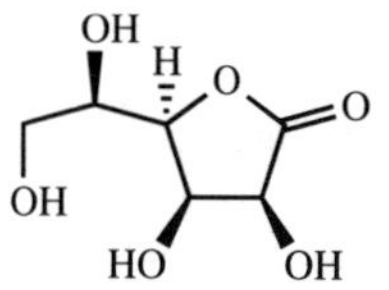

**4.06**
[26301-79-1][3]
D-Mannose(1)

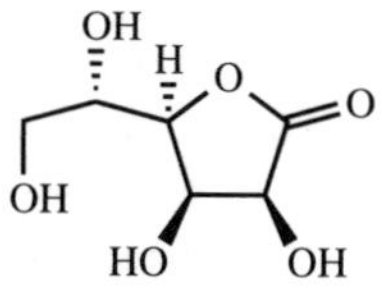

**4.07**
[1128-23-0][4]
L-Ascorbic acid(1)

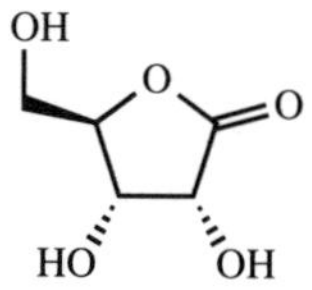

**4.08**
[3327-63-7][5,6]
D-Ribose(1)

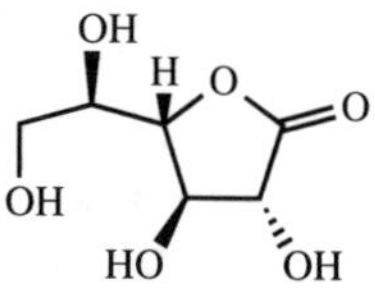

**4.09**
[2782-07-2][5,6]
D-Galactose(1)

## BUILDING BLOCK **4.10**

Name: D-Galactonic acid, 6-Bromo-6-deoxy-γ-lactone triacetate [69617-68-1]

### Synthesis:

(a) $Br_2$, $CaCO_3$, 90-100%[6]

(b) HBr-HOAc, 63%[13]

D-Galactose

### Related Structures:

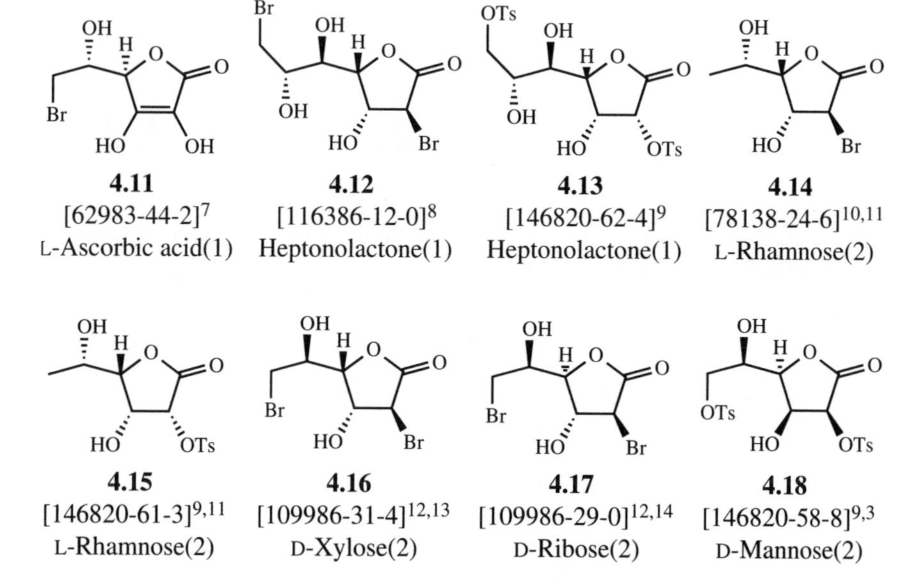

**4.11** [62983-44-2][7] L-Ascorbic acid(1)

**4.12** [116386-12-0][8] Heptonolactone(1)

**4.13** [146820-62-4][9] Heptonolactone(1)

**4.14** [78138-24-6][10,11] L-Rhamnose(2)

**4.15** [146820-61-3][9,11] L-Rhamnose(2)

**4.16** [109986-31-4][12,13] D-Xylose(2)

**4.17** [109986-29-0][12,14] D-Ribose(2)

**4.18** [146820-58-8][9,3] D-Mannose(2)

## BUILDING BLOCK **4.19**

Name: 2(3*H*)-Furanone, 3-Bromodihydro-4-hydroxy-(3*S*-*cis*)-[117858-88-5]

### Synthesis:

L-Ascorbic acid

(a) $H_2O_2$ , $CaCO_3$ , 84%[2]

(b) HBr-HOAc , 73%[15]

### Related Structures:

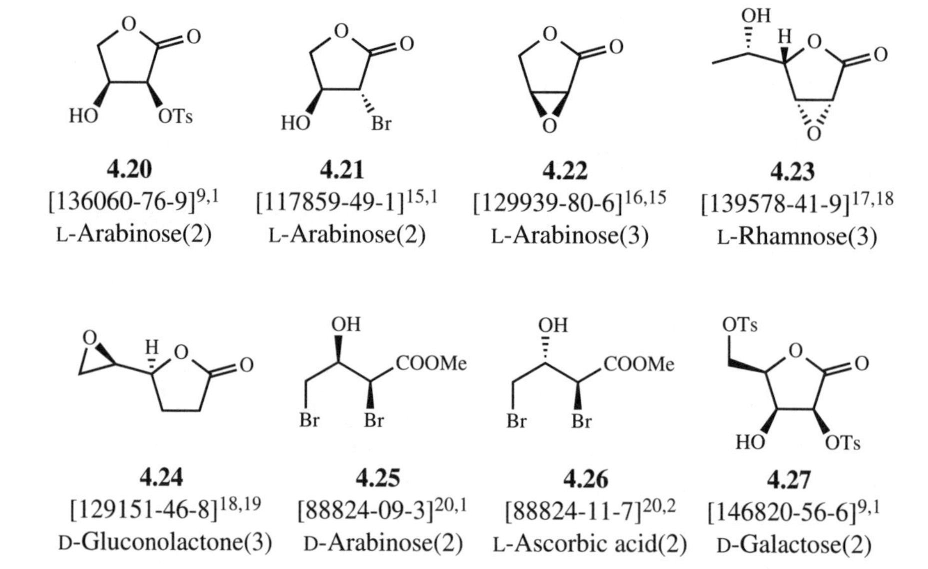

## BUILDING BLOCK **4.28**

OBz
H
O
O
Br
$CH_3OCH_2O$
Br

Name: D-Mannonic acid, 2,6-Dibromo-2,6-dideoxy-3-*O*-methoxy-methyl-γ-lactone 5-benzoate [98120-40-2]

### Synthesis:

OH
O
O
HO
OH
OH

D-Glucono-1,5-lactone

(a) HBr-HOAc, 38-44%[7]

(b) BzCl, pyr., 65%[21]

(c) $(CH_3O)_2CH_2$, $P_2O_5$, 90%

OBz
H
O
O
Br
$CH_3OCH_2O$
Br

### Related Structures:

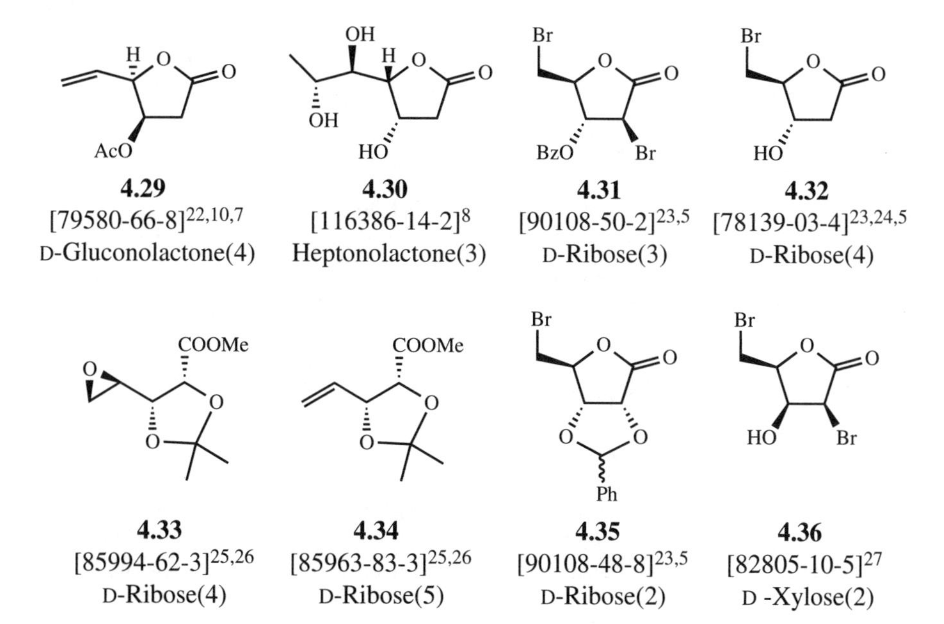

## BUILDING BLOCK **4.37**

Name: D-Lyxonic acid, 2,3-Anhydro-5-bromo-5-deoxy-γ-lactone [129870-06-0]

### Synthesis:

D-Galactose

(a) $O_2$, KOH, 68%[1]

(b) HBr-HOAc, 45-80%[24]

(c) $K_2CO_3$, 93%[16]

### Related Structures:

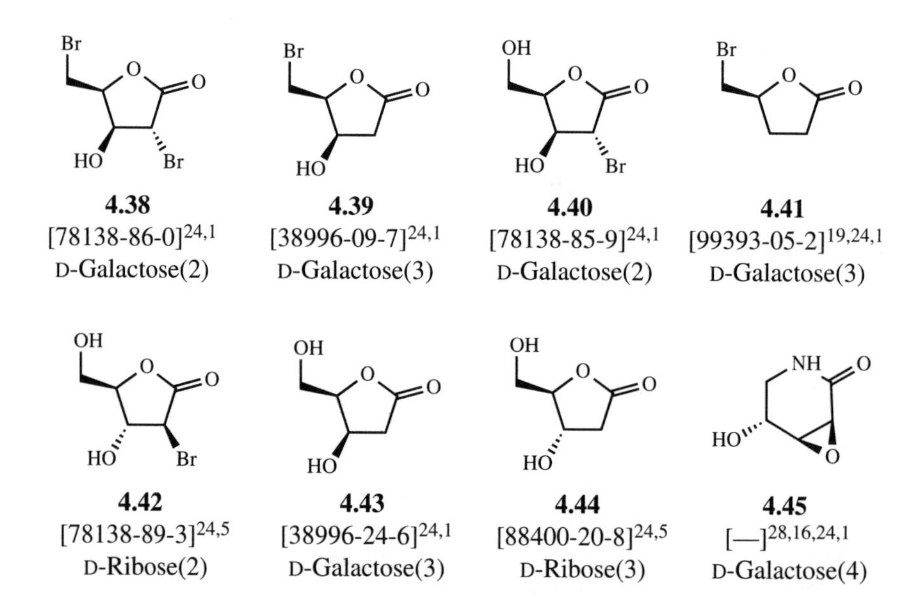

## BUILDING BLOCK **4.46**

Name: 2(3*H*)-Furanone, 5-(2-Bromo-1-hydroxyethyl-dihydro-(5*S*,5′*S*)-
[107855-21-0]

### Synthesis:

D-Glucono-1,5-lactone

(a) HBr-HOAc, 38-44%[7]

(b) $H_2$ , Pd/C, 71%[19]

### Related Structures:

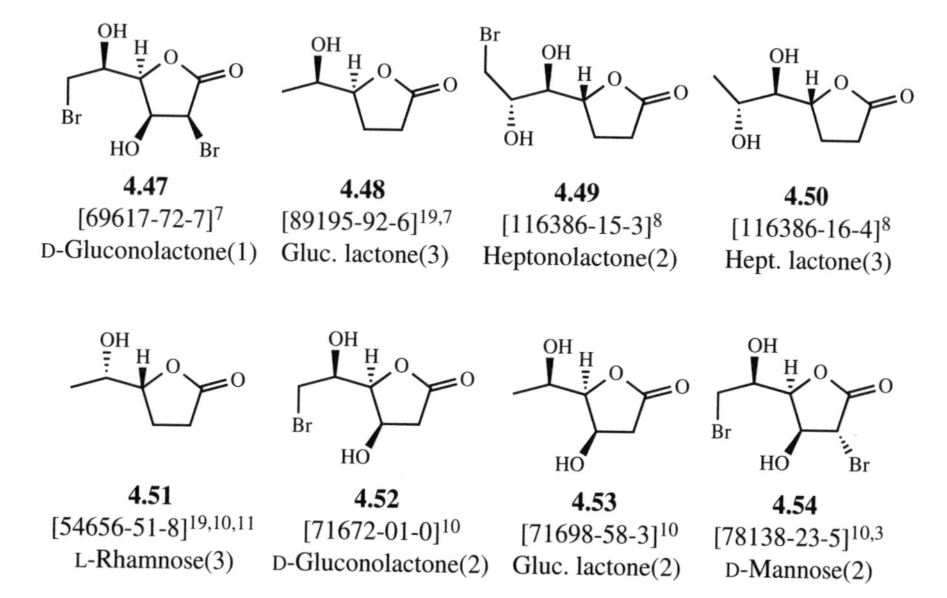

## BUILDING BLOCK **4.55**

OTr

O

O

Name: 2(5*H*)-Furanone, 5-[(Triphenylmethoxy)methyl]-[78598-79-9]

### Synthesis:

D-Ribose

(a) $Br_2$, $CaCO_3$, 90-100%[5]

(b) TrCl, pyr., 84%[29]

(c) S=C(Imidazole)$_2$,79%[29]

(d) Raney Ni, 73%[29]

### Related Structures:

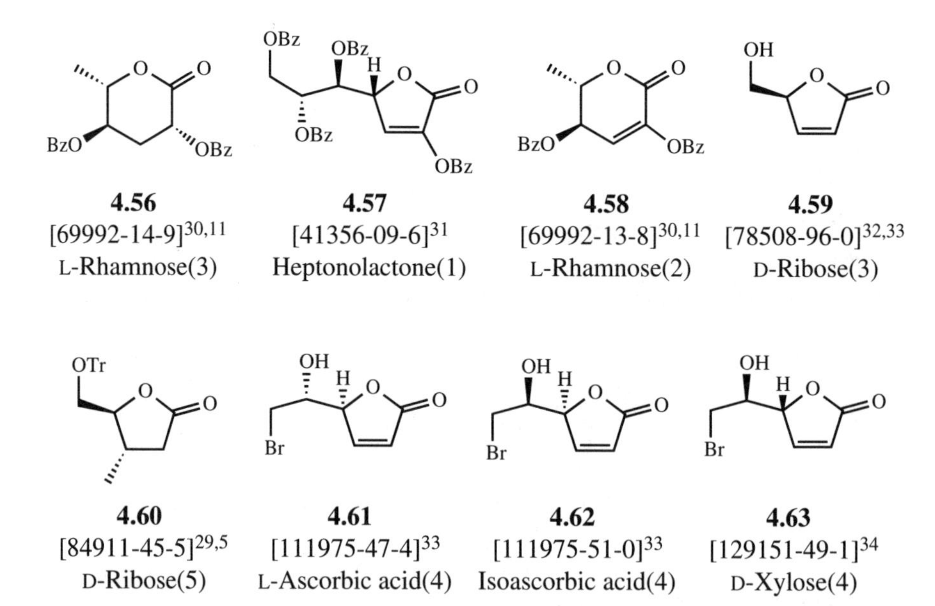

# BUILDING BLOCK **4.64**

OH
H O
O
OH
OH

Name: Hexonic acid, 3-Deoxy-D-*xylo*-γ-lactone [6936-66-9]

## Synthesis:

D-Galactose

(a) $Br_2$ , $CaCO_3$, 89%[6]

(b) $Ac_2O$ , $H^+$, 90%[22]

(c) $H_2$ , Pd/C, 99%[22]

(d) NaOMe , 80%[22]

## Related Structures:

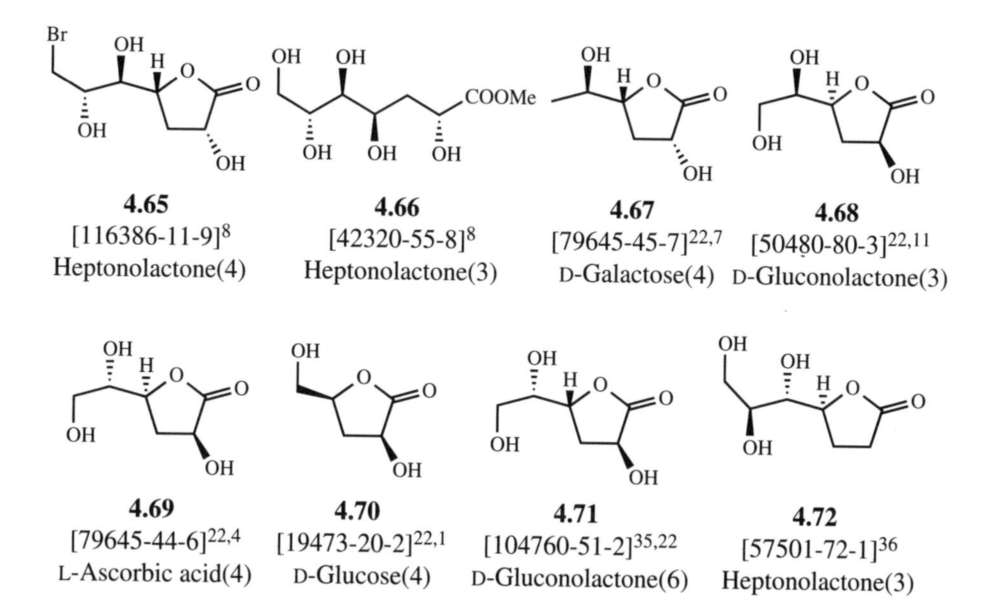

## BUILDING BLOCK **4.73**

Name: L-Gulonic acid, 2,3-*O*-(Methyl)ethylidene-γ-lactone [94840-08-1]

### Synthesis:

L-Ascorbic acid

(a) $H_2$, Pd/C , 99%[4]

(b) $Me_2C(OMe)_2$,TsOH, 79%[37]

(c) $H^+$, 79%[37]

### Related Structures:

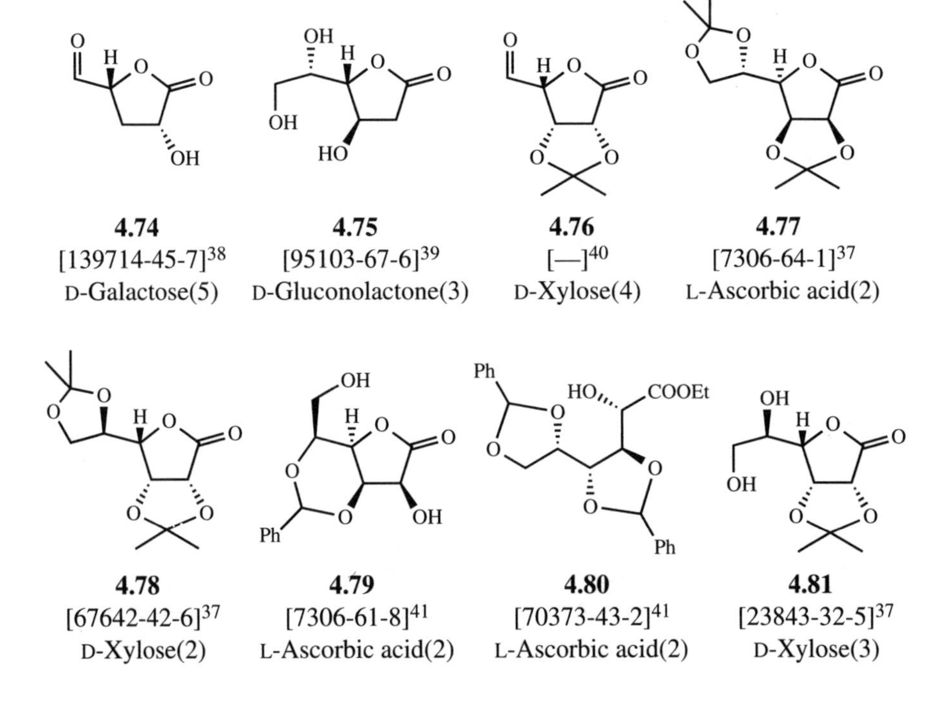

## BUILDING BLOCK **4.82**

Name: D-Ribonic acid, 2,3-*O*-(1-Methylethylidene)-γ-lactone [30725-00-9]

### Synthesis:

D-Ribose

(a) $Br_2$, $CaCO_3$, 40-50%[5]

(b) $Me_2CO$, $H^+$, 74%[26]

### Related Structures:

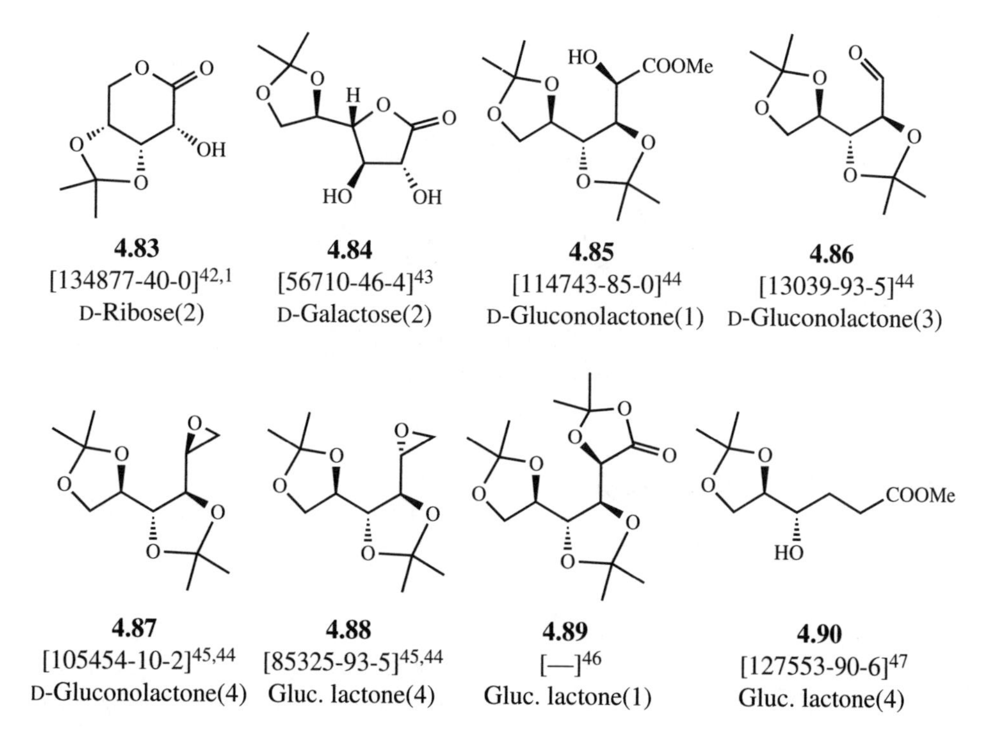

**4.83** [134877-40-0][42,1] D-Ribose(2)

**4.84** [56710-46-4][43] D-Galactose(2)

**4.85** [114743-85-0][44] D-Gluconolactone(1)

**4.86** [13039-93-5][44] D-Gluconolactone(3)

**4.87** [105454-10-2][45,44] D-Gluconolactone(4)

**4.88** [85325-93-5][45,44] Gluc. lactone(4)

**4.89** [—][46] Gluc. lactone(1)

**4.90** [127553-90-6][47] Gluc. lactone(4)

## BUILDING BLOCK **4.91**

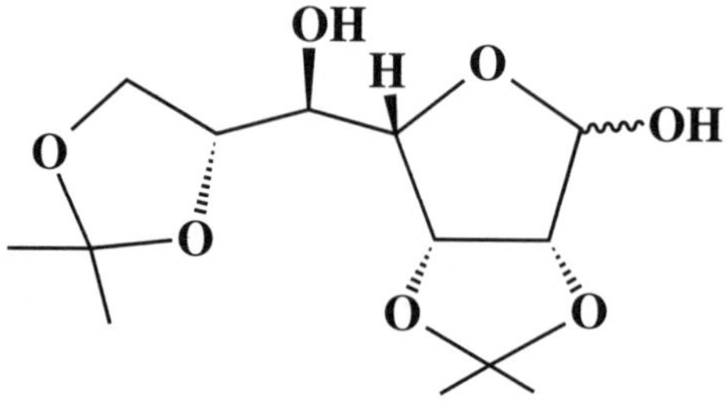

Name: D-*glycero*-D-*gulo*-Heptofuranose, 2,3:6,7-Bis-*O*-(1-methylethylidene)-
[69008-91-9]

### Synthesis:

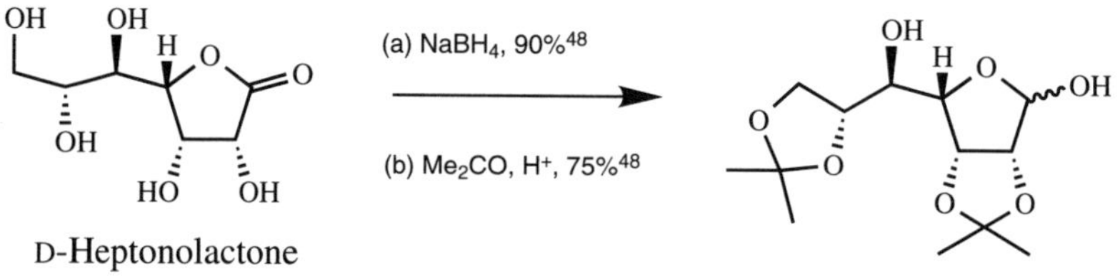

D-Heptonolactone

(a) $NaBH_4$, 90%[48]

(b) $Me_2CO$, $H^+$, 75%[48]

### Related Structures:

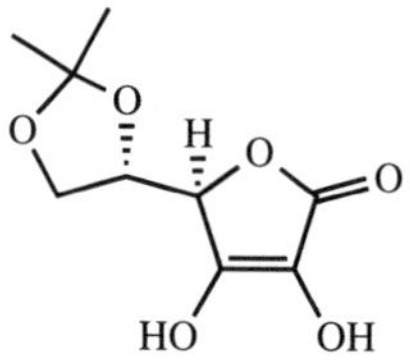

**4.92**
[15042-01-0][49]
L-Ascorbic acid(1)

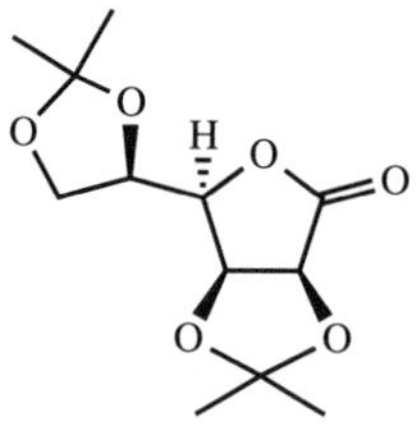

**4.93**
[14440-56-3][50]
D-Mannose(2)

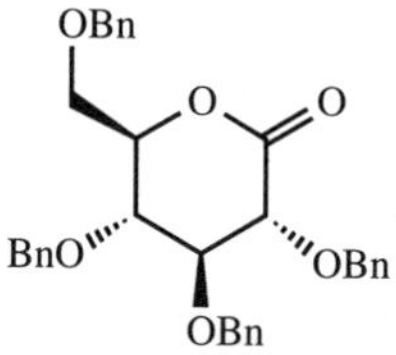

**4.94**
[13096-62-3][51]
D-Glucose(4)

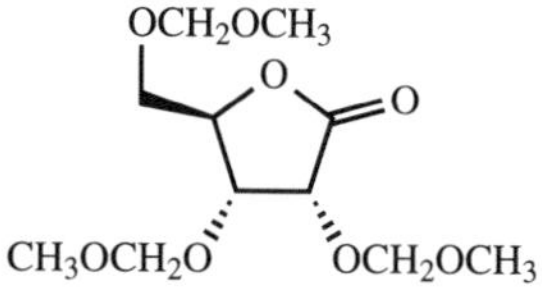

**4.95**
[—][52]
D-Ribose(2)

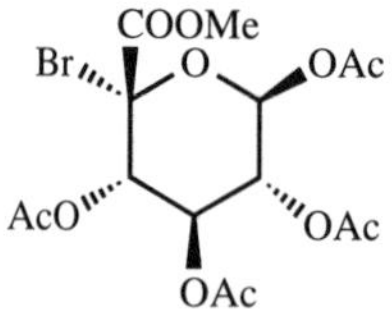

**4.96**
[65615-69-2][53]
Glucuronolactone(3)

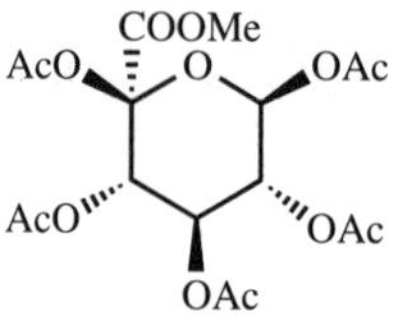

**4.97**
[65615-68-1][54]
Glucuronolactone(4)

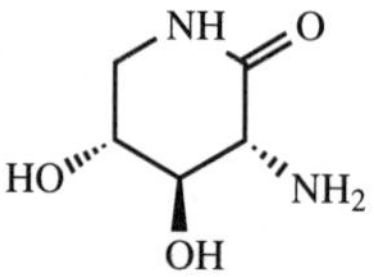

**4.98**
[135030-05-6][55]
Galactose(3)

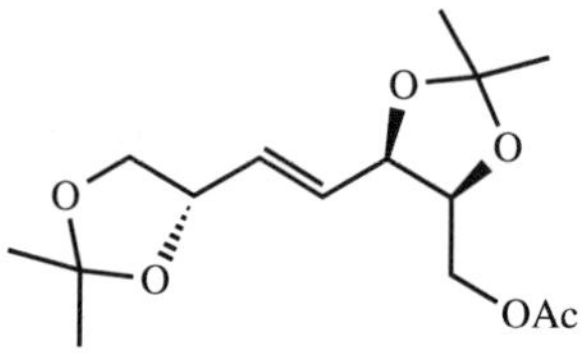

**4.99**
[69008-95-3][48]
Heptonolactone(6)

## BUILDING BLOCK **4.100**

$Bu^tPh_2SiO$

O

O

Name: 1(3*H*)-Isobenzofuranone, 3-[[[(1,1-Dimethylethyl)diphenyl-silyl]oxy]methyl]-3*a*,4,7,7*a*-tetrahydro-[3*S*-(3α,3*a*α,7*a*α)]-[99333-00-3]

### Synthesis:

D-Ribose: HO, OH, OH, OH, O

(a) $Br_2$
(b) $Bu^tPh_2SiCl$, imidazole
(c) S=C(imidazole)$_2$
(d) Raney Ni, 65% (3 steps)
(e) $CH_2$=CH—CH=$CH_2$, $AlCl_3$[56]
75-80%

Product: $Bu^tPh_2SiO$, O, O

### Related Structures:

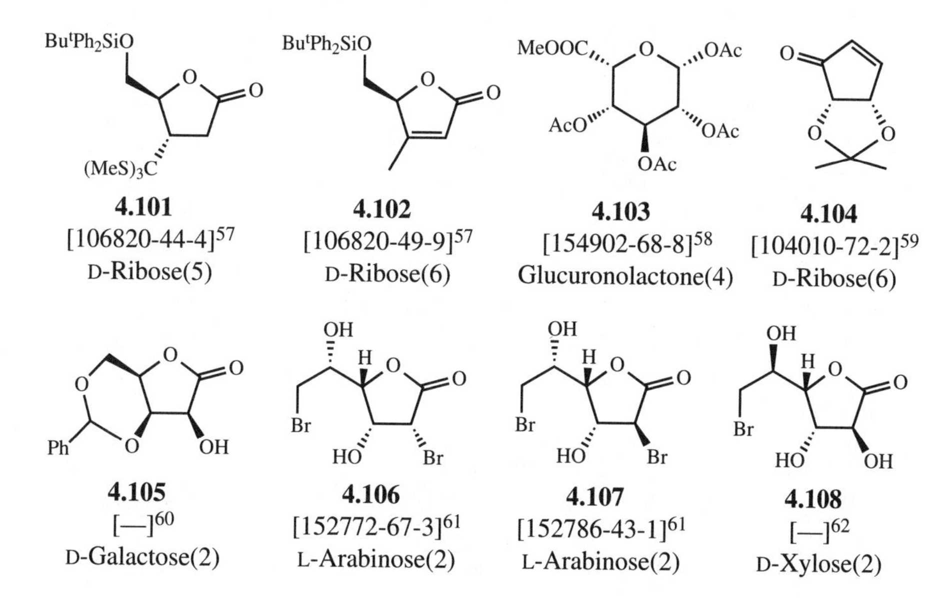


**4.101** [106820-44-4][57] D-Ribose(5)

**4.102** [106820-49-9][57] D-Ribose(6)

**4.103** [154902-68-8][58] Glucuronolactone(4)

**4.104** [104010-72-2][59] D-Ribose(6)

**4.105** [—][60] D-Galactose(2)

**4.106** [152772-67-3][61] L-Arabinose(2)

**4.107** [152786-43-1][61] L-Arabinose(2)

**4.108** [—][62] D-Xylose(2)

## REFERENCES


1. W. J. Humphlett. *Carbohydr. Res.* **1967** *4:*157–164.
2. H. S. Isbell and H. L. Frush. *Carbohydr. Res.* **1979** *72:*301–304.
3. H. S. Isbell. *Methods Carbohydr. Chem.* **1963** *2:*13–14.
4. G. C. Andrews, T. C. Crawford, and B. E. Bacon. *J. Org. Chem.* **1981** *46:*2976–2977.
5. H. S. Isbell. *Methods Carbohydr. Chem.* **1963** *2:*13–14.
6. H. S. Isbell and H. L. Frush. *Bur. Standards J. Research* **1931** *6:*1145–1152.
7. K. Bock, I. Lundt, and C. Pedersen. *Carbohydr. Res.* **1979** *68:*313–319.
8. K. Bock, I. Lundt, C. Pedersen, and R. Sonnichsen. *Carbohydr. Res.* **1988** *174:*331–340.
9. R. Madsen and I. Lundt. *Synthesis* **1992** 1129–1132.
10. K. Bock, I. Lundt, and C. Pedersen. *Carbohydr. Res.* **1981** *90:*7–16.
11. H. S. Isbell and H. L. Frush. *Bur. Stand. J. Res.* **1933** *11:*649–664.
12. K. Bock, I. Lundt, C. Pedersen, and S. Refn. *Acta Chem. Scand.* **1986** *B40:*740–744.
13. J.V. Karubinos. *Org. Synth. Coll.* **1963** *4:*506–508.
14. F. L. Humoller. *Methods Carbohydr. Chem.* **1962** *1:*102–104.
15. M. Bols and I. Lundt. *Acta Chem. Scand.* **1988** *B42:*67–74.
16. M. Bols and I. Lundt. *Acta Chem. Scand.* **1990** *44:*252–256.
17. M. Bols, I. Lundt, and E. R. Ottosen. *Carbohydr. Res.* **1991** *222:*141–149.
18. I. Lundt and C. Pedersen. *Synthesis* **1992** 669–672.
19. I. Lundt and C. Pedersen. *Synthesis* **1986** 1052–1054.
20. K. Bock, I. Lundt, and C. Pedersen. *Acta Chem. Scand.* **1983** *B37:*341–344.
21. R. J. Ferrier and P. C. Tyler. *Carbohydr. Res.* **1985** *136:*249–258.
22. K. Bock, I. Lundt, and C. Pedersen. *Acta Chem. Scand.* **1981** *B35:*155–162.
23. S.Y. Chen and M. M. Joullie. *J. Org. Chem.* **1984** *49:*2168–2174.
24. K. Bock, I. Lundt, and C. Pedersen. *Carbohydr. Res.* **1981** *90:*17–26.
25. R.W. Hoffman and W. Ladner. *Chem. Ber.* **1983** *116:*1631–1642.
26. L. Hough, J. K. N. Jones, and D. L. Mitchell. *Can. J. Chem.* **1958** *36:*1720–1728.
27. K. Bock, I. Lundt, and C. Pedersen. *Carbohydr. Res.* **1982** *104:*79–85.
28. T. M. Jespersen, M. Bols, M. Sierks, and T. Skrydstrup. *Tetrahedron* **1994** *50:*13449–13460.
29. R. E. Ireland, R. C. Andersen, R. Baboud, B. J. Fitzsimmons, G. J. McGarvey, S. Thaisrivongs, and C. S. Wilcox. *J. Am. Chem. Soc.* **1983** *105:*1988–2006.
30. O. J. Varela, A. F. Cirelli, R. M. de Lederkremer. *Carbohydr. Res.* **1979** *70:*27–35.
31. L. F. Sala, A. F. Cirelli, and R. M. de Lederkremer. *Carbohydr. Res.* **1980** *78:*61–66.
32. P. Camps, J. Cardellach, J. Font, R. M. Ortuno, and O. Ponsati. *Tetrahedron* **1982** *38:*2395–2402.
33. J. A. J. M. Vekemans, G. A. M. Franken, C.W. Dapperens, E. F. Godefroi, and G. J. F. Chittenden. *J. Org. Chem.* **1988** *53:*627–633.
34. J. A. J. M. Vekemans, C.W. Dapperens, R. Claessen, A. M. J. Koten, E. F. Godefroi, and G. J. F. Chittenden. *J. Org. Chem.* **1990** *55:*5336–5344.
35. K. Bock, I. Lundt, and C. Pedersen. *Acta Chem. Scand.* **1986** *B40:*163–171.

36. K. Bock, I. Lundt, and C. Pedersen. *Carbohydr. Res.* **1988** *179:*87–96.
37. G.W. J. Fleet, N. G. Ramsden, and D. R. Witty. *Tetrahedron* **1989** *45:*319–326.
38. M. Bols, I. Lundt, and C. Pedersen. *Tetrahedron* **1992** *48:*319–324.
39. K. Bock, I. Lundt, and C. Pedersen. *Acta Chem. Scand.* **1984** *B38:*555–561.
40. R. K. Hulyalkar and J. K. N. Jones. *Can. J. Chem.* **1963** *41:*1898–1904.
41. T. C. Crawford and R. Breitenbach. *J. Chem. Soc. Chem. Commun.* **1979** 388–389.
42. H. Zinner, H. Voigt, and J. Voigt. *Carbohydr. Res.* **1968** *7:*38–55.
43. J. A. J. M. Vekemans, R. G. M. de Bruin, R. C. H. M. Caris, A. J. P. M. Kokx, J. J. H. G. Konings, E. F. Godefroi, and G. J. F. Chittenden. *J. Org. Chem.* **1987** *52:*1093–1099.
44. H. Regeling, E. de Rouville, and G. J. F. Chittenden. *Recl. Trav. Chim. Pays-Bas* **1987** *106:*461–464.
45. H. Regeling and G. J. F. Chittenden. *Carbohydr. Res.* **1989** *190:*313–316.
46. J.W.W. Morgan and M. L. Wolfrom. *J. Am. Chem. Soc.* **1956** *78:*2496–2497.
47. H. Regeling and G. J. F. Chittenden. *Carbohydr. Res.* **1990** *205:*261–268.
48. G. Stork, T. Takahashi, I. Kawamoto, and T. Suzuki. *J. Am. Chem. Soc.* **1978** *100:*8272–8273.
49. M. E. Jung and T. J. Shaw. *J. Am. Chem. Soc.* **1980** *102:*6304–6311.
50. T. F. Tam and B. Fraser-Reid. *J. Org. Chem.* **1980** *45:*1344–1346.
51. B. Rajnikanth and R. Seshadri. *Tetrahedron Lett.* **1989** *30:*755–758.
52. M. Bols, H. Gruppe, T. M. Jespersen, and W. A. Szarek. *Carbohydr. Res.* **1994** *253:*195–206.
53. R. J. Ferrier and P. C. Tyler. *J. Chem. Soc. Chem. Commun.* **1978** 1019–1020.
54. R. J. Ferrier and R. H. Furneaux. *J. Chem. Soc. Perkin I* **1977** 1996–2000.
55. M. Bols and I. Lundt. *Acta Chem. Scand.* **1991** *45:*280–284.
56. M. G. B. Drew, J. Mann, and A. Thomas. *J. Chem. Soc. Perkin I* **1986** 2279–2285.
57. S. Hanessian and P. J. Murray. *Tetrahedron* **1987** *43:*5055–5072.
58. D. Medakovic. *Carbohydr. Res.* **1994** *253:*299–300.
59. H. Ohrui, M. Konno, and M. Meguro. *Agric. Biol. Chem.* **1987** *51:*625–626.
60. S.-Y. Han, M. M. Joullie, V.V. Fokin, and N. A. Petasis. *Tetrahedron: Asymmetry* **1994** *5:*2535–2562.
61. I. Lundt and R. Madsen. *Synthesis* **1993** 714–720.
62. I. Lundt and H. Frank. *Tetrahedron* **1994** *50:*13285–13298.

## BUILDING BLOCK **5.01**

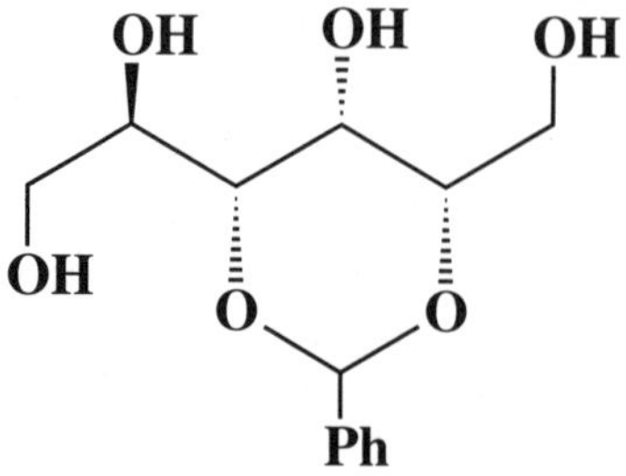

Name: D-Glucitol, 2,4-*O*-(Phenylmethylene)-[77340-95-5]

### Synthesis:

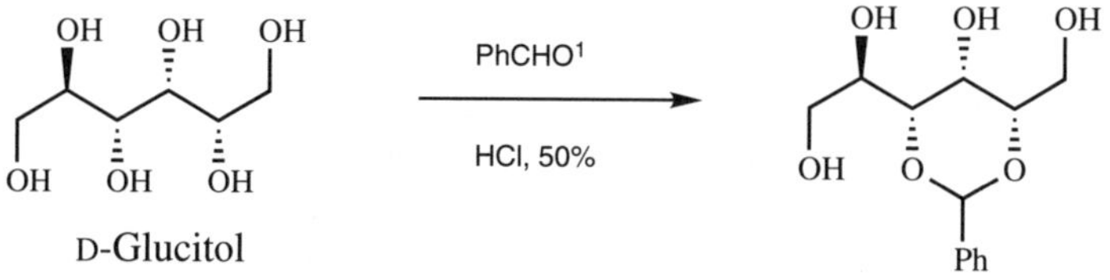

D-Glucitol

PhCHO[1]

HCl, 50%

### Related Structures:

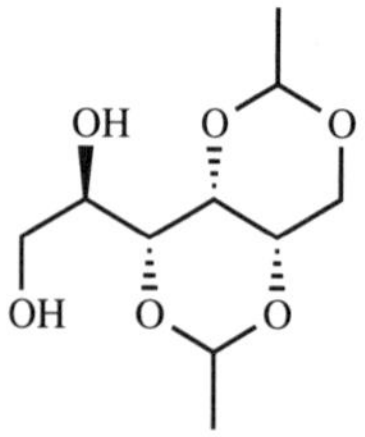

**5.02**
[50895-31-3][2]
D-Glucitol(1)

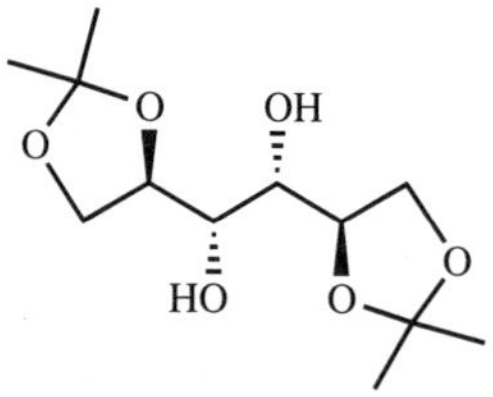

**5.03**
[1707-77-3][3]
D-Mannitol(1)

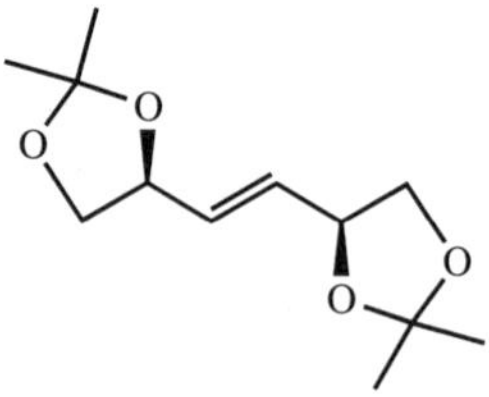

**5.04**
[3427-24-5][4]
D-Mannitol(3)

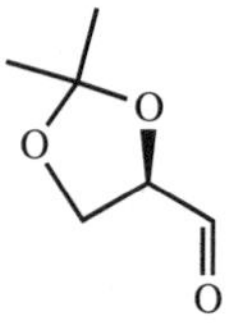

**5.05**
[15186-48-8][5]
D-Mannitol(2)

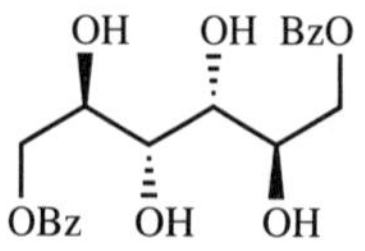

**5.06**
[7226-27-9][6]
D-Mannitol(1)

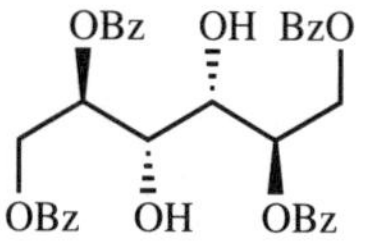

**5.07**
[20934-11-1][7]
D-Mannitol(3)

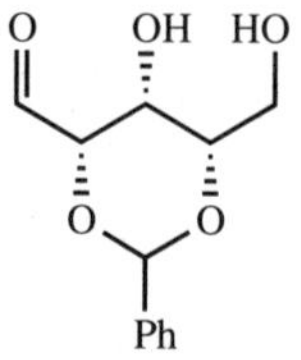

**5.08**
[30608-02-7][8]
D-Glucitol(2)

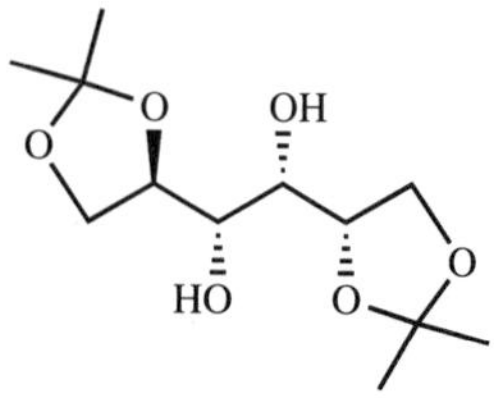

**5.09**
[53735-98-1][9]
D-Glucitol(1)

## BUILDING BLOCK **5.10**

Name: D-Mannitol, 3,4-*O*-(1-Methylethylidene)-[3969-84-4]

### Synthesis:

D-Mannitol

(a) Acetone, $H^+$, 75%[10]

(b) $H_3O^+$, 80%[10]

### Related Structures:

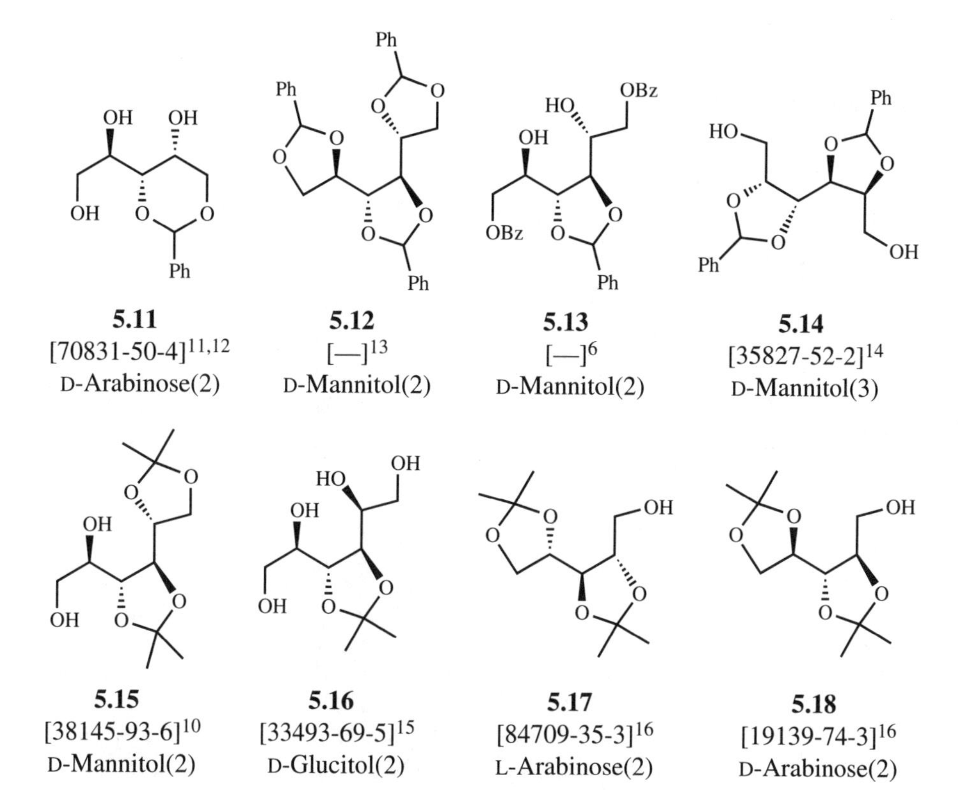

**5.11** [70831-50-4][11,12] D-Arabinose(2)

**5.12** [—][13] D-Mannitol(2)

**5.13** [—][6] D-Mannitol(2)

**5.14** [35827-52-2][14] D-Mannitol(3)

**5.15** [38145-93-6][10] D-Mannitol(2)

**5.16** [33493-69-5][15] D-Glucitol(2)

**5.17** [84709-35-3][16] L-Arabinose(2)

**5.18** [19139-74-3][16] D-Arabinose(2)

## BUILDING BLOCK **5.19**

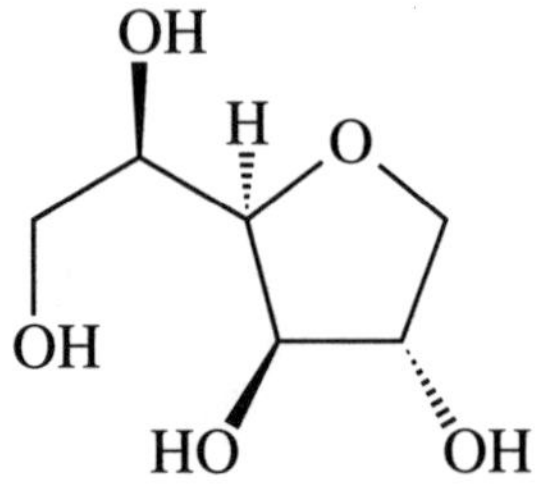

OH
H
O
OH
HO OH

Name: D-Glucitol, 1,4-Anhydro-[27299-12-3]

### Synthesis:

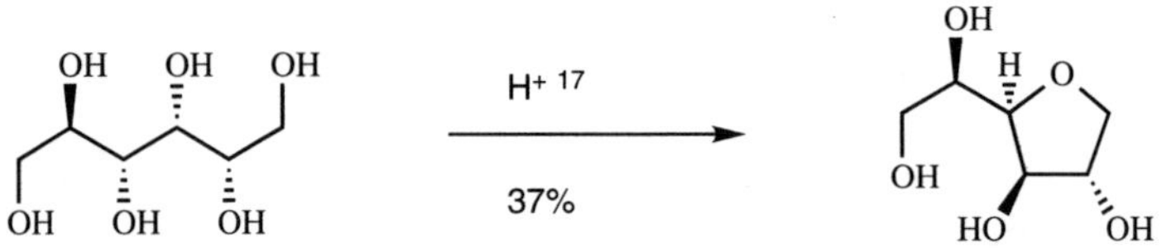

OH OH OH
OH OH OH
D-Glucitol

$H^+$ [17]

37%

### Related Structures:

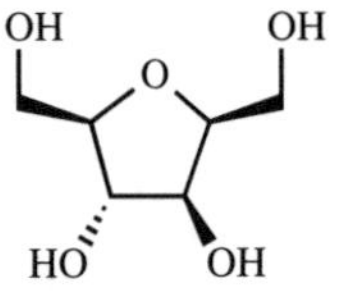

**5.20**
[27826-73-9][18]
D-Mannitol(3)

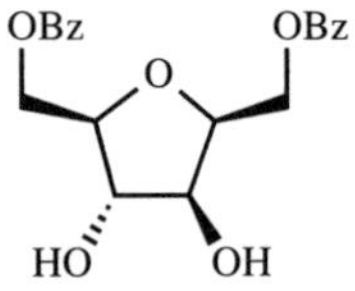

**5.21**
[22144-41-8][18]
D-Mannitol(2)

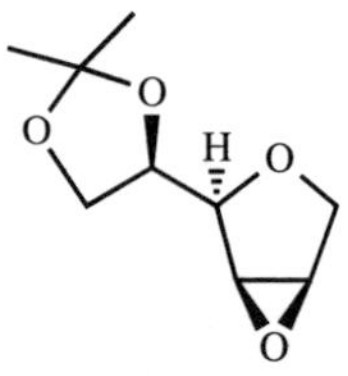

**5.22**
[55730-76-2][19]
D-Glucitol(4)

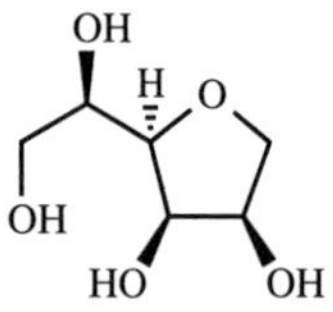

**5.23**
[7726-97-8][20]
D-Mannitol(1)

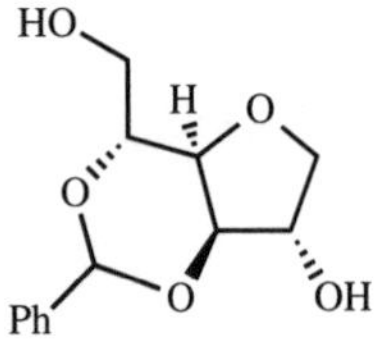

**5.24**
[121250-37-1][21]
D-Glucitol(2)

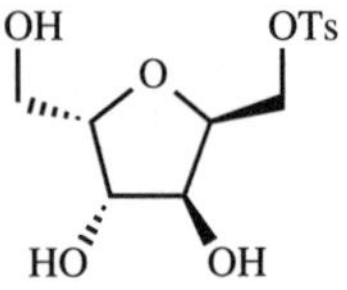

**5.25**
[32445-71-9][22]
D-Glucitol(4)

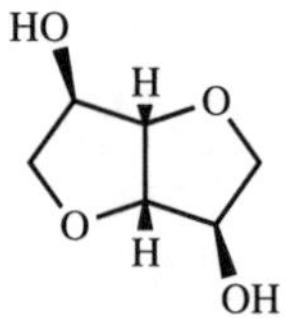

**5.26**
[28948-16-5][23]
D-Mannitol(4)

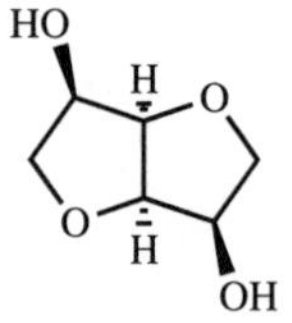

**5.27**
[641-74-7][24]
D-Mannitol(1)

# BUILDING BLOCK **5.28**

Name: L-Glucitol, 1,4:3,6-Anhydro-[124508-14-1]

## Synthesis:

D-Mannose

(a) $Br_2$[25]
(b) HBr-HOAc, 38-44%[26]
(c) $NaBH_4$, 95%[27]
(d) KOH, 74%[27]

## Related Structures:

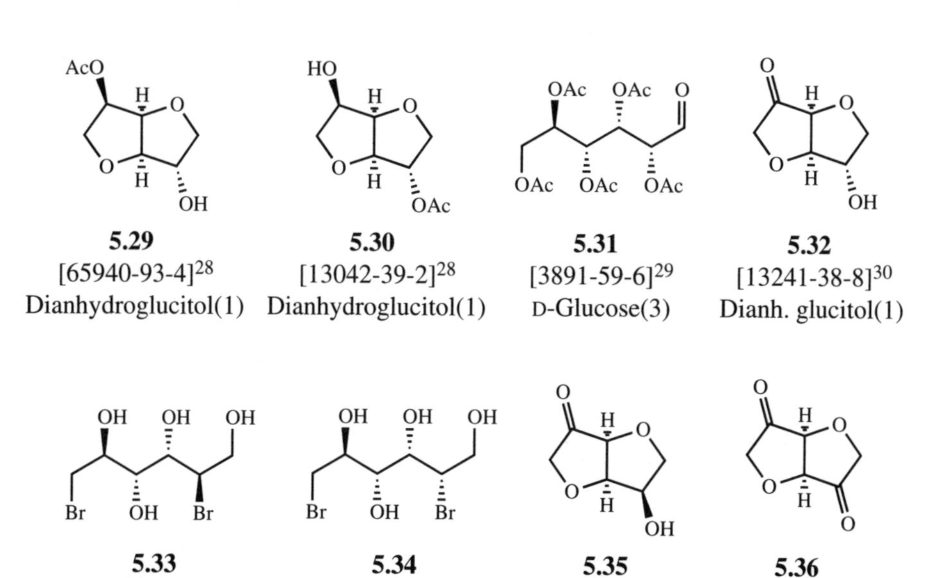

**5.29** [65940-93-4][28] Dianhydroglucitol(1)

**5.30** [13042-39-2][28] Dianhydroglucitol(1)

**5.31** [3891-59-6][29] D-Glucose(3)

**5.32** [13241-38-8][30] Dianh. glucitol(1)

**5.33** [71672-05-4][31] D-Gluconolactone(2)

**5.34** [124379-09-5][27] D-Mannose(3)

**5.35** [13241-40-2][32] D-Mannitol(2)

**5.36** [13241-36-6][32] D-Mannitol(2)

## BUILDING BLOCK **5.37**

Name: D-Pentitol, 2,3:4,5-Bis-*O*-(1-methylethylidene)-1-*C*-phenyl-[—]

### Synthesis:

D-Arabinose

(a) EtSH, $H^+$, 73%[33]
(b) Acetone, $H^+$, 91%[34]
(c) HgO, $HgCl_2$, 71%[35]
(d) PhMgBr, 54%[36]

### Related Structures:

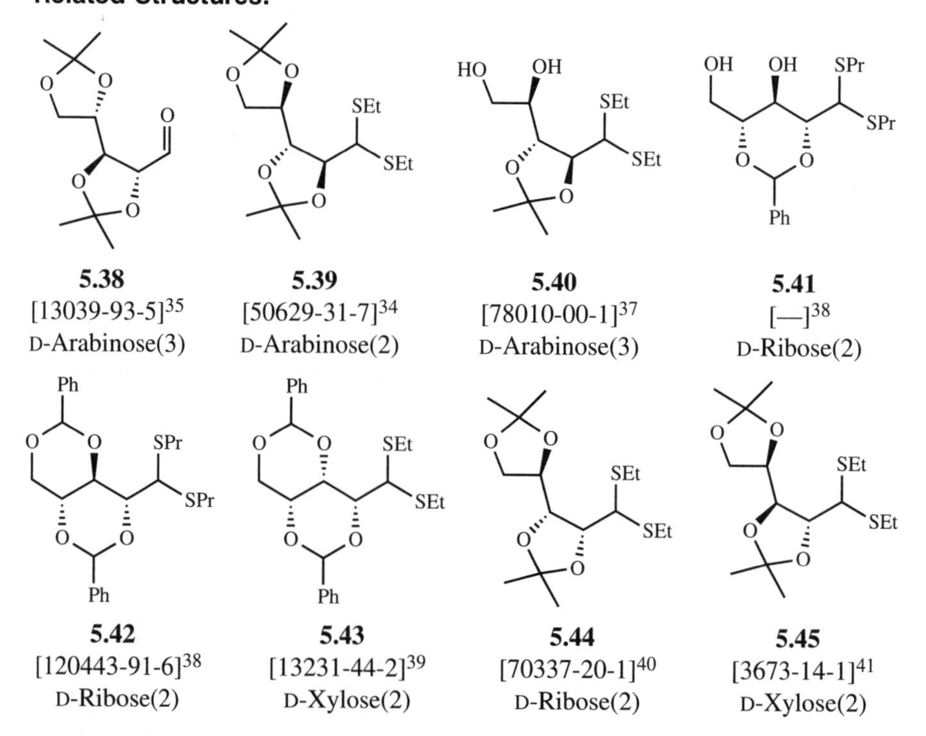

## BUILDING BLOCK **5.46**

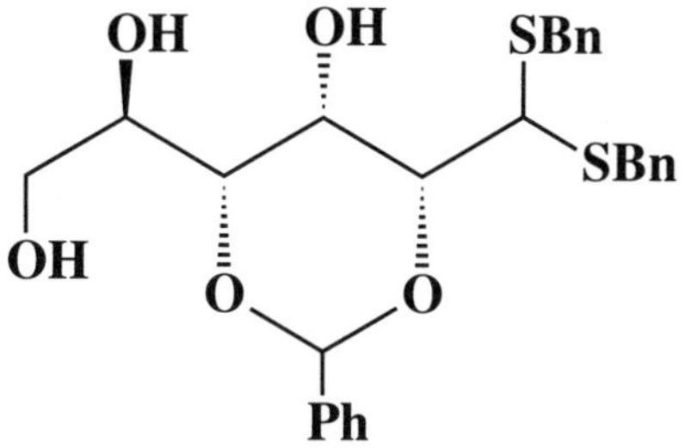

Name: D-Glucose, 2,4-*O*-Phenylmethylene-bis(phenylmethyl)-mercaptal [—]

### Synthesis:

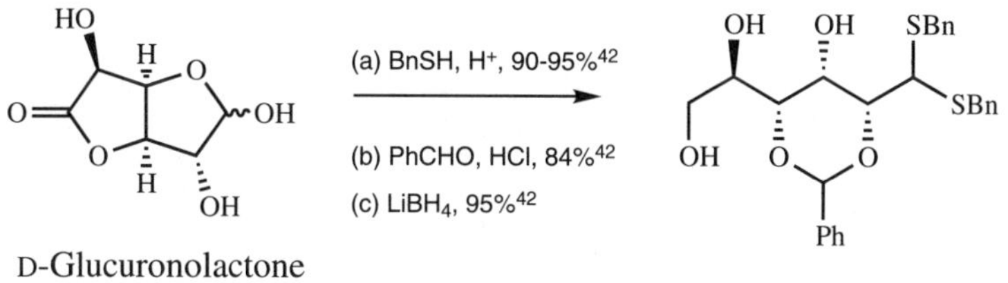

D-Glucuronolactone

(a) BnSH, $H^+$, 90-95%[42]

(b) PhCHO, HCl, 84%[42]

(c) $LiBH_4$, 95%[42]

### Related Structures:

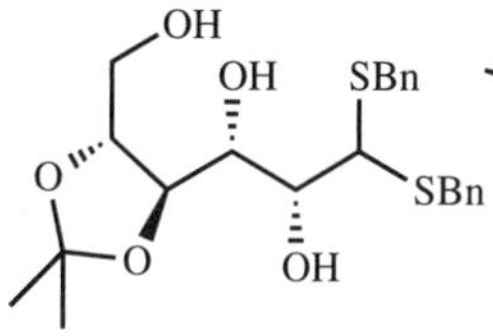

**5.47**
[—][43]
D-Galactose(2)

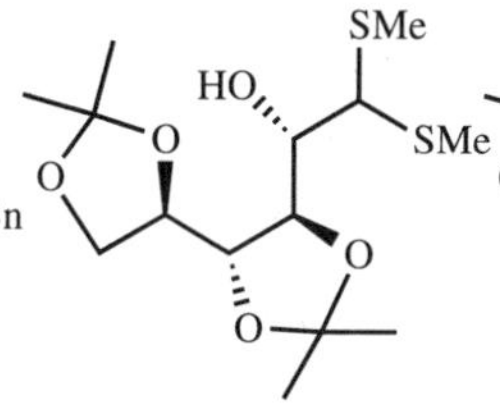

**5.48**
[—][44]
D-Mannose(2)

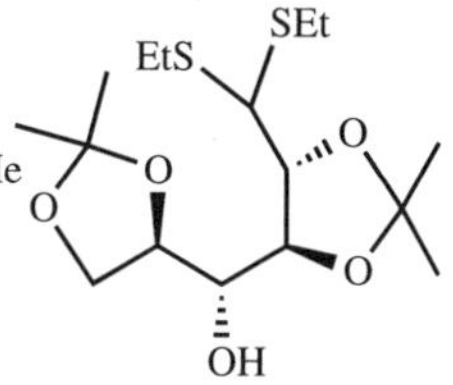

**5.49**
[4258-02-0][45]
D-Glucose(2)

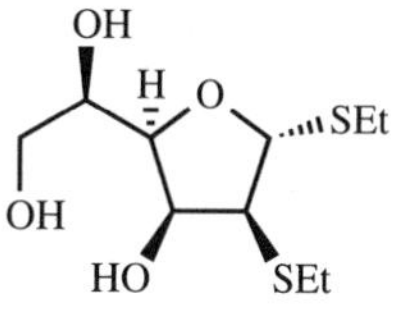

**5.50**
[30085-91-7][46]
D-Glucosamine(2)

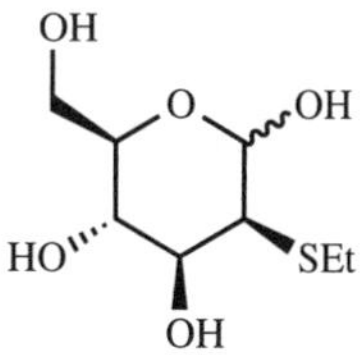

**5.51**
[15356-42-0][47]
D-Glucosamine(2)

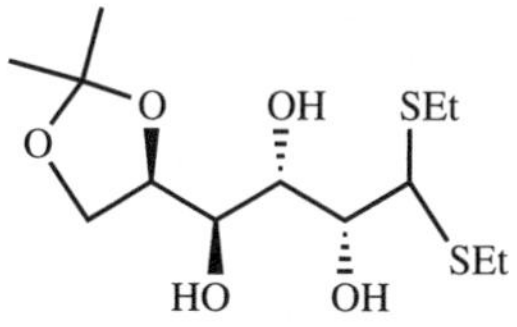

**5.52**
[99773-30-5][48]
D-Galactose(2)

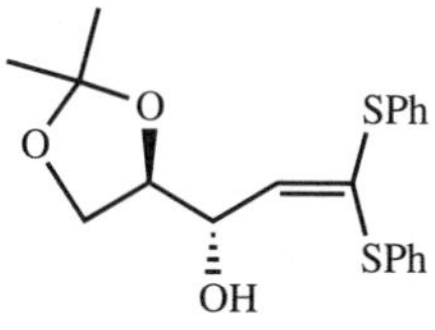

**5.53**
[28697-89-4][34]
D-Arabinose(3)

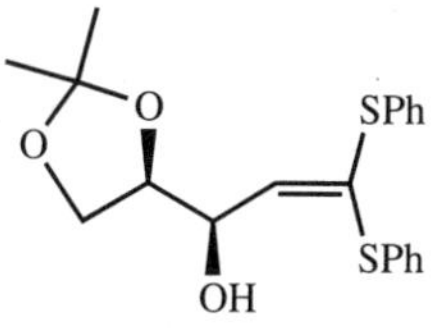

**5.54**
[37107-88-3][41]
D-Xylose(3)

## BUILDING BLOCK **5.55**

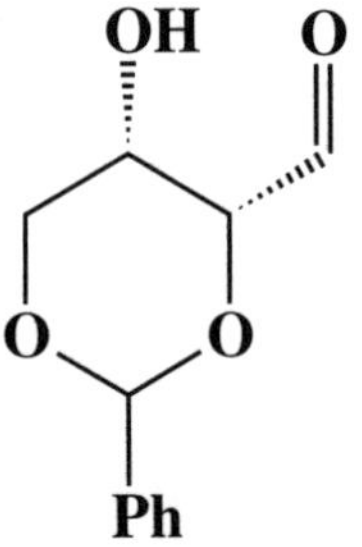

Name: 1,3-Dioxane-4-carboxaldehyde, 5-Hydroxy-2-phenyl-[2*R*-(2$\alpha$,4$\alpha$,5$\alpha$)]
[—]

### Synthesis:

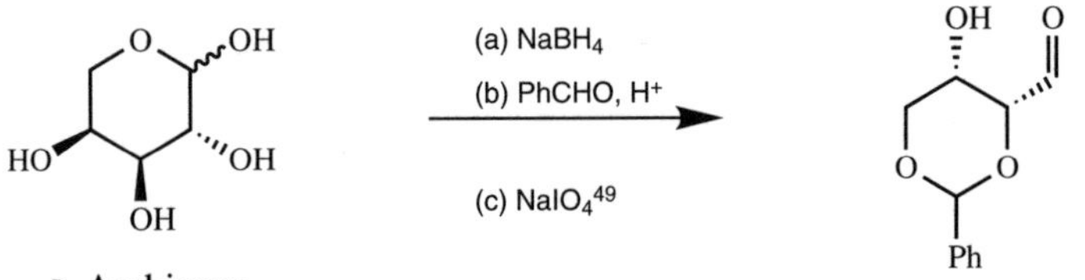

L-Arabinose

(a) $NaBH_4$
(b) PhCHO, $H^+$
(c) $NaIO_4$[49]

### Related Structures:

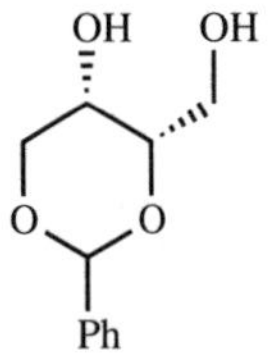

**5.56**
[—][49]
L-Arabinose(4)

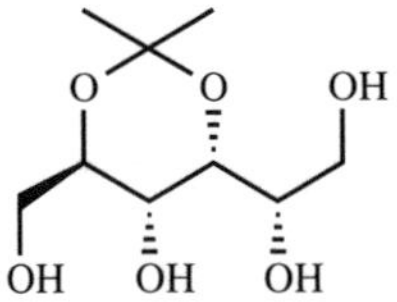

**5.57**
[144664-16-4][50]
D-Glucitol(5)

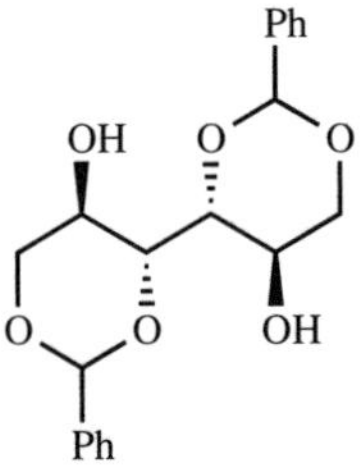

**5.58**
[28224-73-9][51]
D-Mannitol(1)

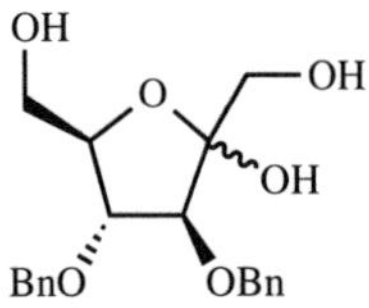

**5.59**
[—][52]
D-Mannitol(4)

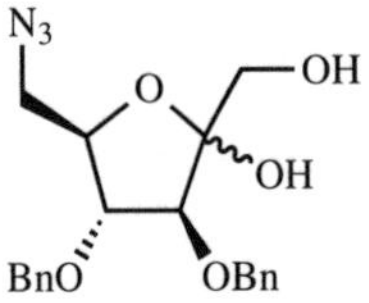

**5.60**
[—][52]
D-Mannitol(6)

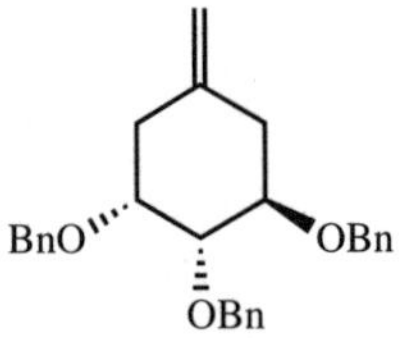

**5.61**
[34298-67-4][53]
D-Arabinose(6)

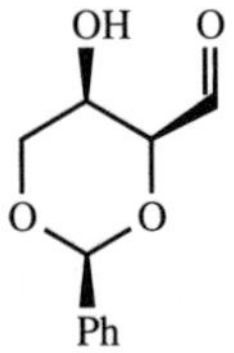

**5.62**
[99274-32-5][49]
D-Arabinose(3)

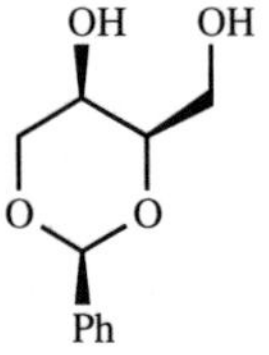

**5.63**
[81577-58-4][49]
D-Arabinose(4)

## BUILDING BLOCK **5.64**

Name: L-Iditol, 1,5-Dibromo-1,5-dideoxy-[152884-04-3]

### Synthesis:

L-Ascorbic acid

(a) $H_2$, Pd/C, 99%
(b) HBr-HOAc, 90%
(c) $NaBH_4$, 51%[54]

### Related Structures:

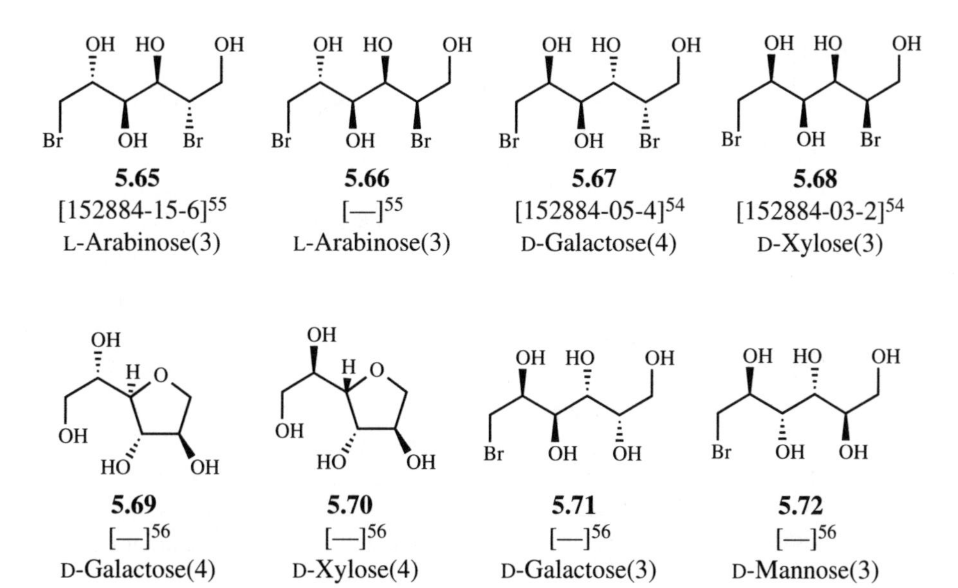

**5.65** [152884-15-6][55] L-Arabinose(3)

**5.66** [—][55] L-Arabinose(3)

**5.67** [152884-05-4][54] D-Galactose(4)

**5.68** [152884-03-2][54] D-Xylose(3)

**5.69** [—][56] D-Galactose(4)

**5.70** [—][56] D-Xylose(4)

**5.71** [—][56] D-Galactose(3)

**5.72** [—][56] D-Mannose(3)

## BUILDING BLOCK **5.73**

OH HO OH
$NO_2$ O O
Ph

Name: D-Glucitol, 6-deoxy-6-nitro-2,4-*O*-(phenylmethylene)-[78124-22-8]

### Synthesis:

OH HO OH
OH OH OH
D-Glucitol

(a) PhCHO, $H^+$, 58%
(b) $NaIO_4$
(c) $CH_3NO_2$
NaOMe
50% (2 steps)[57]

OH HO OH
$NO_2$ O O
Ph

### Related Structures:

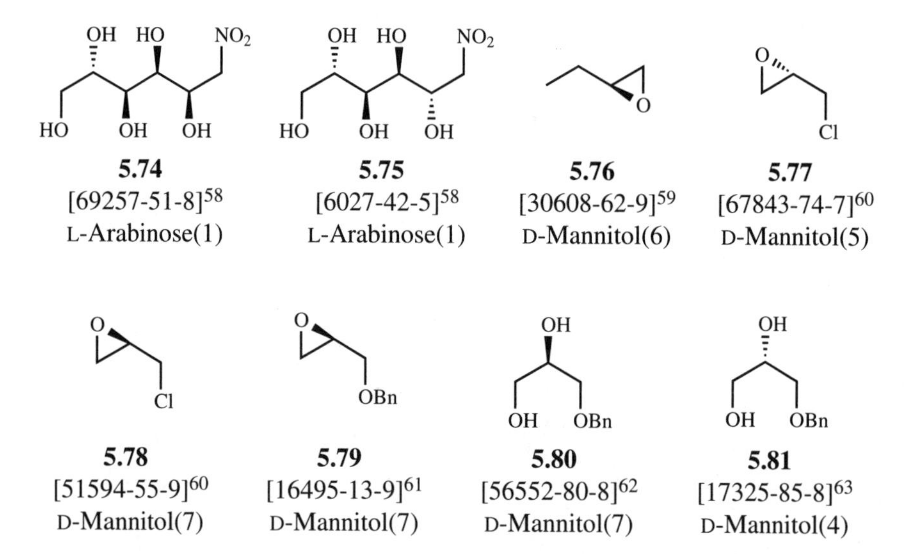

## REFERENCES


1. L. von Vargha. *Chem. Ber.* **1935** *68:*18–24.
2. R. C. Hockett and F. C. Schaefer. *J. Am. Chem. Soc.* **1947** *69:*849–851.
3. G. J. F. Chittenden. *Carbohydr. Res.* **1980** *87:*219–226.
4. C. E. Ballou. *Methods Carbohydr. Chem.* **1972** *6:*393–398.
5. J. Mann, N. K. Partlett, and A. Thomas. *J. Chem. Res. (S)* **1987** 369.
6. H. Ohle, H. Erlbach, H. Hepp, and G. Toussaint. *Chem. Ber.* **1929** *62:*2982–2990.
7. J.W. van Cleve. *Carbohydr. Res.* **1982** *106:*170–173.
8. R. N. Ray. *J. Ind. Chem. Soc.* **1987** *64:*371.
9. G. J. F. Chittenden. *Carbohydr. Res.* **1982** *108:*81–87.
10. L. F. Wiggins. *J. Chem. Soc.* **1946** 13–14.
11. E. Fisher. *Chem. Ber.* **1894** *27:*1524–1537.
12. O. Ruff. *Chem. Ber.* **1899** *32:*550–560.
13. J.W. Pette. *Chem. Ber.* **1931** *64:*1567–1568.
14. W. T. Haskins, R. M. Hann, and C. S. Hudson. *J. Am. Chem. Soc.* **1943** *65:*1419–1422.
15. E. J. Bourne, G. P. McSweeney, M. Stacey, and L. F. Wiggins. *J. Chem. Soc.* **1946** 13–14.
16. A. Holy. *Collect. Czech. Chem. Commun.* **1982** *47:*2786–2805.
17. S. Solzberg, R. M. Goepp, Jr., and W. Freudenberg. *J. Am. Chem. Soc.* **1946** *68:*919–921.
18. T. A.W. Koerner, Jr., E. S. Younathan, A.-L. E. Ashour, and R. J. Voll. *J. Biol. Chem.* **1974** *249:*5749–5754.
19. S. Hanessian, P. Dextraze, A. Fougerousse, and Y. Guindon. *Tetrahedron. Lett.* **1974** *46:*3983–3986.
20. A. B. Foster and W. G. Overend. *J. Chem. Soc.* **1951** 680–684.
21. V. G. Bashford and L. F. Wiggins. *Nature* **1950** *165:*566–567.
22. L. von Vargha. *Chem. Ber.* **1935** *68:*1377–1384.
23. D. R. Hicks and B. Fraser-Reid. *Can. J. Chem.* **1974** *52:*3367–3372.
24. J. Feldmann, H. Koebernick, and H. U. Woelk. DE 3,041,673, **1980**; *Chem. Abstr.* **1982** *97:*163,415.
25. H. S. Isbell. *Methods Carbohydr. Chem.* **1963** *2:*13–14.
26. K. Bock, I. Lundt, and C. Pedersen. *Carbohydr. Res.* **1979** *68:*313–319.
27. K. Bock, I. M. Castilla, I. Lundt, and C. Pedersen. *Acta Chem. Scand.* **1989** *43:*264–268.
28. P. Stoss, P. Merrath, and G. Schlüter. *Synthesis* **1987** 174–176.
29. M. L. Wolfrom and A. Thompson. *Methods Carbohydr. Chem.* **1963** *2:*427–430.
30. F. Jacquet, L. Rigal, and A. Gaset. *J. Chem. Technol. Biotechnol.* **1990** *48:*493–506.
31. K. Bock, I. Lundt, and C. Pedersen. *Carbohydr. Res.* **1981** *90:*7–16.
32. K. Heyns, W.-P. Trautwein, and H. Paulsen. *Chem. Ber.* **1963** *96:*3195–3199.
33. M. L. Wolfrom, D. I. Weisblat, W. H. Zophy, and S.W. Waisbrot. *J. Am. Chem. Soc.* **1941** *63:*201–203.

34. D. Horton and J. D. Wander. *Carbohydr. Res.* **1970** *13:*33–47.
35. R. Ramage, A. M. McLeod, and G.W. Rose. *Tetrahedron* **1991** *47:*5625–5636.
36. W. A. Bonner. *J. Am. Chem. Soc.* **1951** *73:*3126–3132.
37. H. Zinner, G. Rembarz, and H. P. Klöcking. *Chem. Ber.* **1957** *90:*2688–2696.
38. D. J. J. Potgieter and D. L. MacDonald. *J. Org. Chem.* **1961** *26:*3934–3938.
39. H. Zinner, B. Richard, M. Blessmann, and M. Schlutt. *Carbohydr. Res.* **1966** *2:*197–203.
40. M. A. Bakhari, A. B. Foster, J. Lehmann, J. M. Webber, and J. H. Westwood. *J. Chem. Soc.* **1963** 2291–2295.
41. B. Berrang, D. Horton, and J. D. Wander. *J. Org. Chem.* **1973** *38:*187–192.
42. H. Zinner, C.-G. Dässler. *Chem. Ber.* **1960** *93:*1597–1608.
43. E. Pascu, S. M. Trister, and J.W. Green. *J. Am. Chem. Soc.* **1939** *61:*2444–2448.
44. E. J. Curtis and J. K. N. Jones. *Can. J. Chem.* **1960** *38:*890–895.
45. T. B. Grindley and G. Wickranage. *Carbohydr. Res.* **1987** *167:*105–121.
46. J. Defaye. *Bull. Soc. Chim. Fr.* **1967** 1101–1103.
47. A. E. El Ashmawy, D. Horton, L. G. Magbanua, and J. M. J. Tronchet. *Carbohydr. Res.* **1968** *6:*299–309.
48. M. L. Wolfrom and G. G. Parekh. *Carbohydr. Res.* **1969** *11:*547–557.
49. A. B. Foster, A. H. Haines, J. Homer, J. Lehmann, and L. F. Thomas. *J. Chem. Soc.* **1961** 5005–5011.
50. A. J. Al-Kadir, N. Bagett, J. M. Webber. *Carbohydr. Res.* **1992** *232:*249–257.
51. H. B. Sinclair. *Carbohydr. Res.* **1970** *12:*150–172.
52. W. Zou and W. A. Szarek. *Carbohydr. Res.* **1994** *254:*25–33.
53. H. J. Bestmann and H. A. Heid. *Angew. Chem. Int. Ed.* **1971** *10:*336–337.
54. I. Lundt and R. Madsen. *Synthesis* **1993** 720–724.
55. I. Lundt and R. Madsen. *Synthesis* **1993** 714–720.
56. I. Lundt and H. Frank. *Tetrahedron* **1994** *50:*13285–13298.
57. R. L. Whistler and J. N. BeMiller. *Methods Carbohydr. Chem.* **1962** *1:*137–139.
58. J. C. Sowden. *Methods Carbohydr. Chem.* **1962** *1:*132–135.
59. U. Schmidt, J. Talbiersky, F. Bartkowiak, and J. Wild. *Angew. Chem. Int. Ed.* **1980** *19:*198–199.
60. J. J. Baldwin, A.W. Raab, K. Mensler, B. H. Arison, and D. E. McClure. *J. Org. Chem.* **1978** *43:*4876–4878.
61. A. K. M. Anisuzzaman and L. N. Owen. *J. Chem. Soc. (C)* **1967** 1021–1026.
62. W. E. M. Lunds and A. Zschocke. *J. Lipid Res.* **1965** *6:*324–325.
63. S. Takeno, E. Goto, M. Hirama, and K. Ogasawara. *Heterocycles* **1981** *16:*381–385.

# BUILDING BLOCK **6.01**

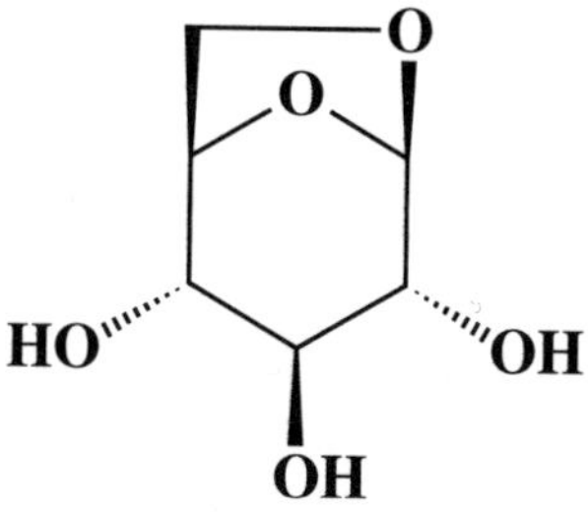

Name: β-D-Glucopyranose, 1,6-Anhydro-[498-07-7]

## Synthesis:

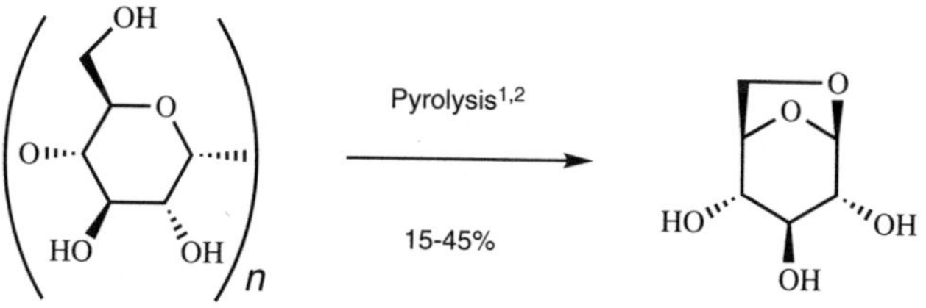

Starch

Pyrolysis[1,2]

15-45%

## Related Structures:

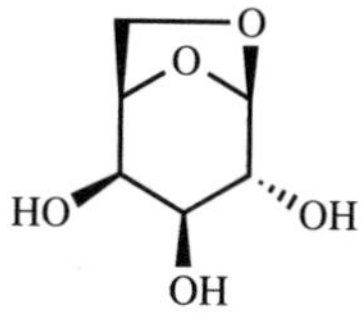

**6.02**
[644-76-8][3]
D-Galactose(3)

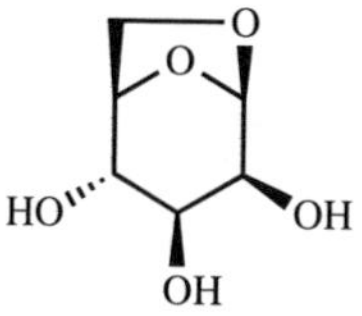

**6.03**
[14168-65-1][4,5]
D-Mannose(3)

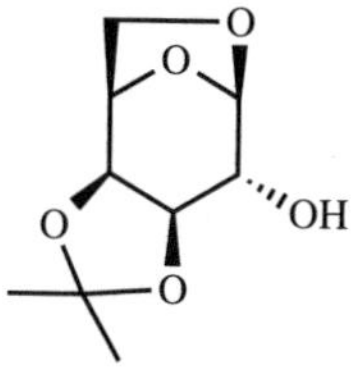

**6.04**
[52579-97-2][6]
Lactose(2)

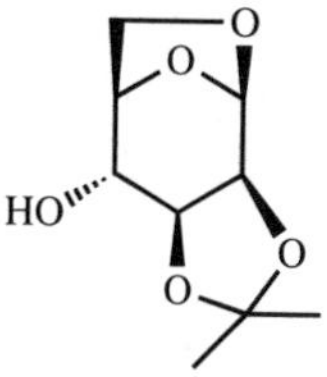

**6.05**
[14440-51-8][4,5]
D-Mannose(2)

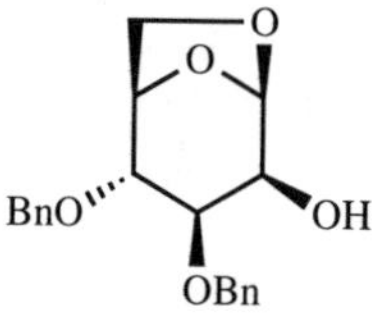

**6.06**
[67227-89-8][7]
D-Mannose(5)

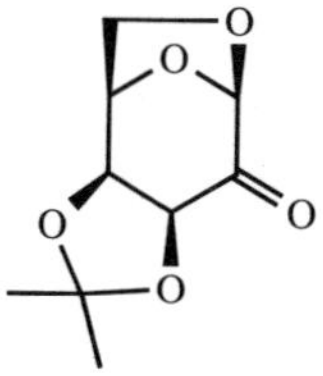

**6.07**
[17073-94-8][8]
Lactose(3)

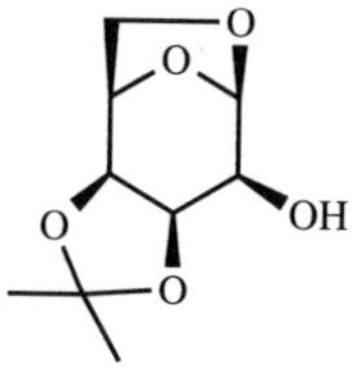

**6.08**
[14278-75-2][8]
Lactose(4)

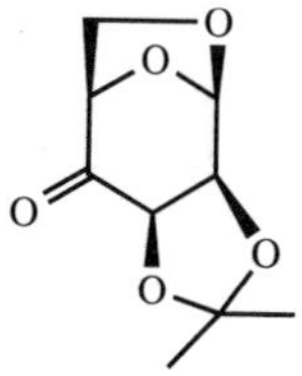

**6.09**
[14440-50-7][9]
D-Mannose(3)

## BUILDING BLOCK **6.10**

Name: β-D-Mannopyranose, 1,6:2,3-Dianhydro-4-*O*-(phenylmethyl)-
[33208-47-8]

### Synthesis:

Starch

(a) pyrolysis, 15-45%[1,2]
(b) TsCl, pyridine[10]
(c) NaOMe, 55% (2 steps)[11]
(d) BnOH, $H^+$, 62%[10]
(e) NaOMe, 89%[12]

### Related Structures:

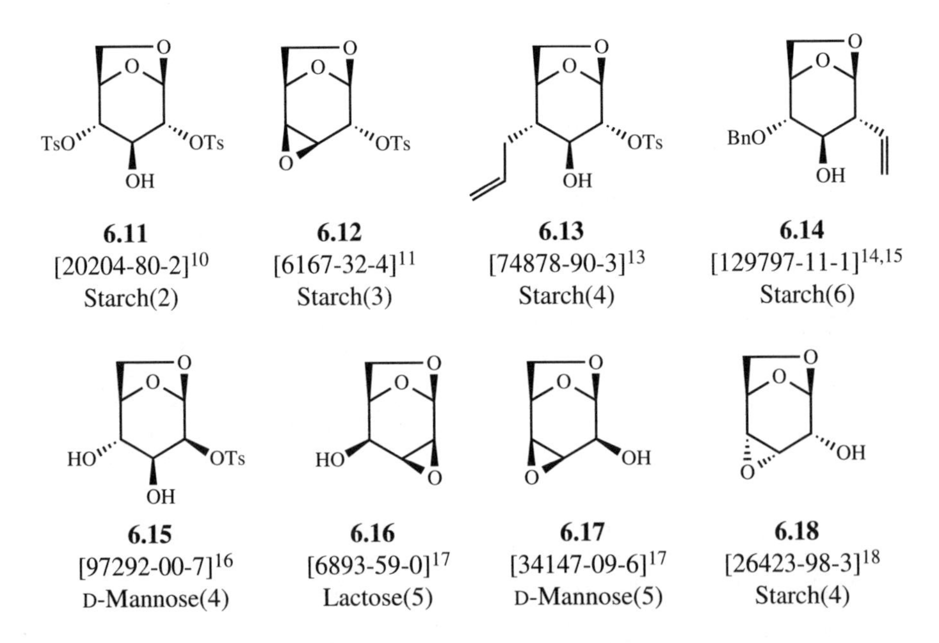

## BUILDING BLOCK **6.19**

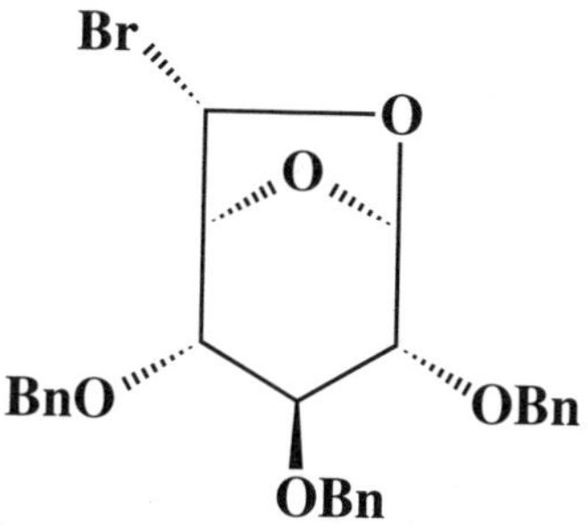

Name: β-D-Glucopyranose, 1,6-Dianhydro-6-*C*-bromo-tribenzoate [74774-22-4]

### Synthesis:

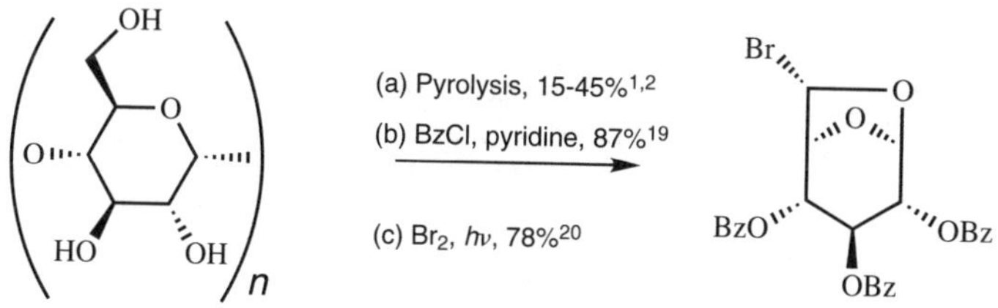

Starch

(a) Pyrolysis, 15-45%[1,2]
(b) BzCl, pyridine, 87%[19]
(c) $Br_2$, $h\nu$, 78%[20]

### Related Structures:

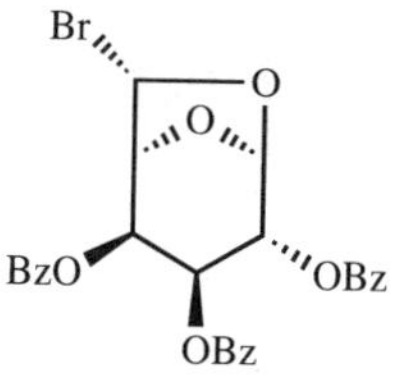

**6.20**
[91876-34-5][21]
Lactose(5)

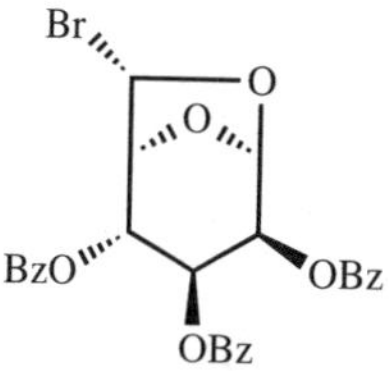

**6.21**
[110567-02-7][22]
D-Mannose(5)

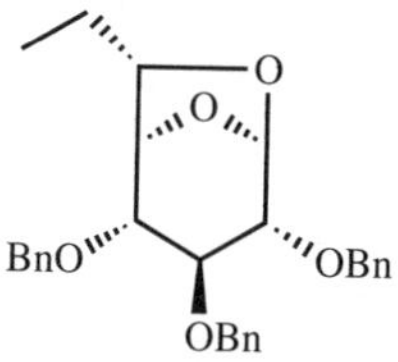

**6.22**
[98853-88-4][23]
Starch(8)

**6.23**
[110808-40-7][24]
Ribose(3)

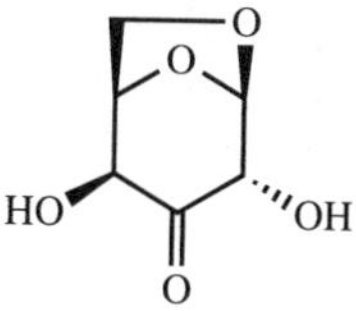

**6.24**
[23094-53-3][25]
Lactose(4)

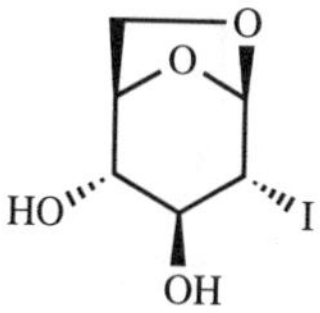

**6.25**
[139437-39-1][26]
D-Glucose(4)

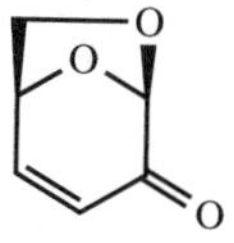

**6.26**
[37112-31-5][27,28]
Cellulose(1)

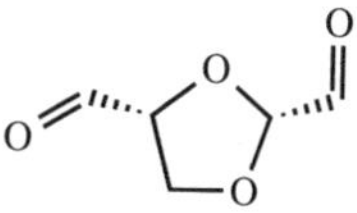

**6.27**
[127903-51-9][29]
Starch(2)

## BUILDING BLOCK **6.28**

O O OTs OH

Name: β-D-*xylo*-Hexopyranose, 1,6-Dianhydro-4-deoxy-2-(4-methylbenzenesulfonate) [23643-30-3]

### Synthesis:

OH O O HO OH n

Starch

(a) Pyrolysis, 15-45%[1,2]

(b) TsCl, pyridine[10]

(c) NaOMe, 55% (2 steps)[11]

(d) $H_2$, Ni, 65-70%[30]

O O OTs OH

### Related Structures:

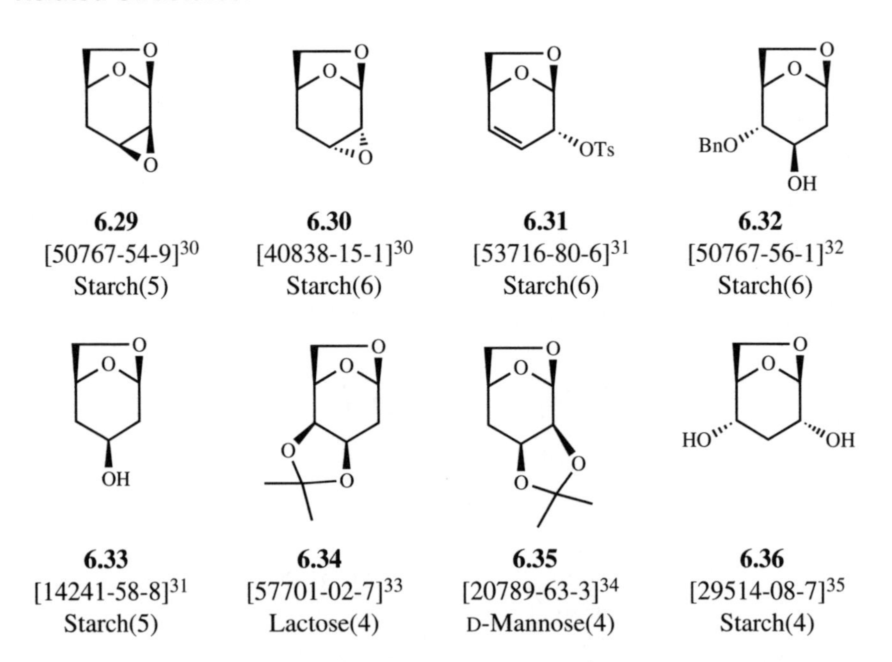

## BUILDING BLOCK **6.37**

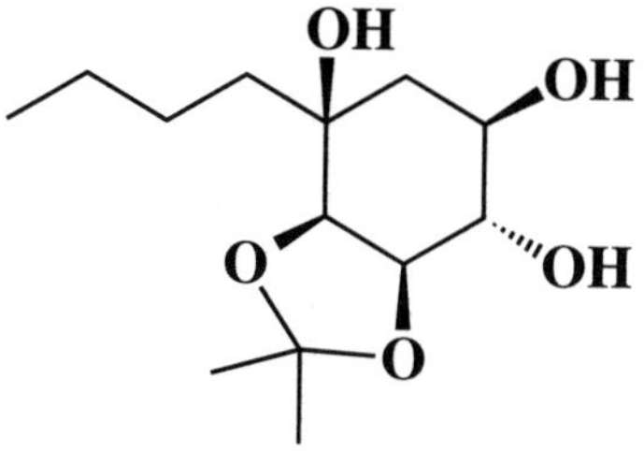

Name: D-*epi*-Inositol, 3-*C*-Butyl-2-deoxy-4,5-*O*-(1-methylethylidene)-
[93366-80-4]

### Synthesis:

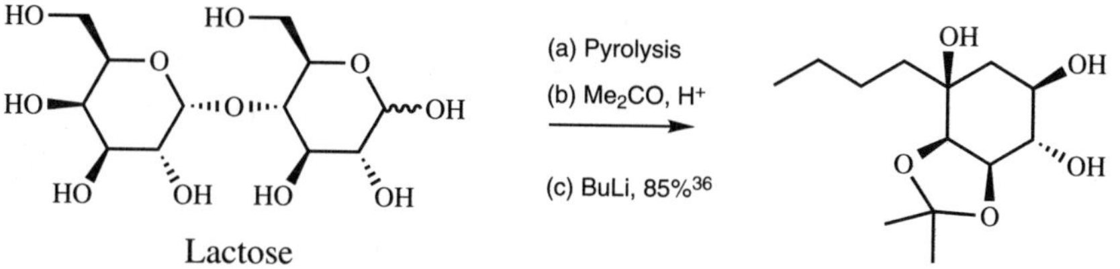

Lactose

(a) Pyrolysis
(b) $Me_2CO$, $H^+$
(c) BuLi, 85%[36]

### Related Structures:

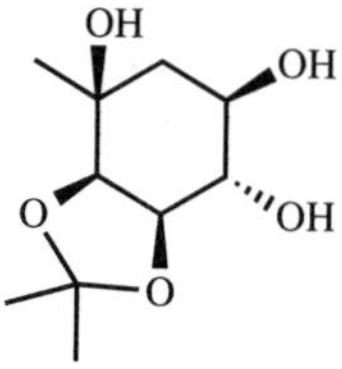

**6.38**
[93366-82-6][36]
Lactose(3)

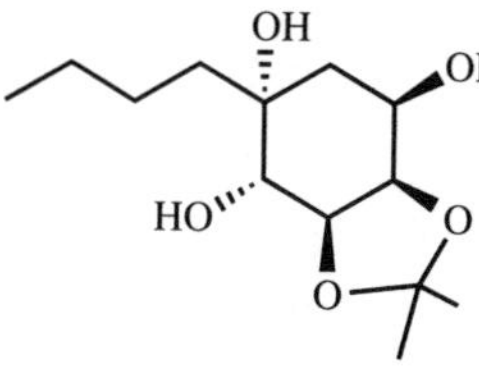

**6.39**
[93366-84-8][36]
D-Mannose(3)

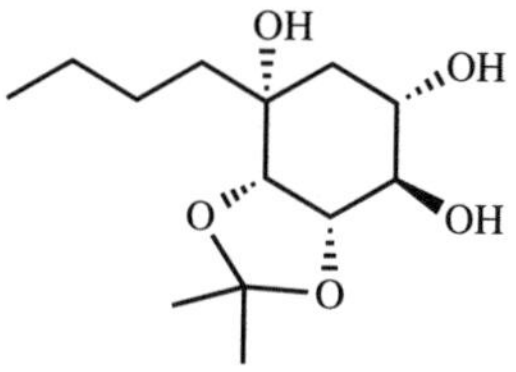

**6.40**
[100760-74-5][37]
D-Glucose(7)

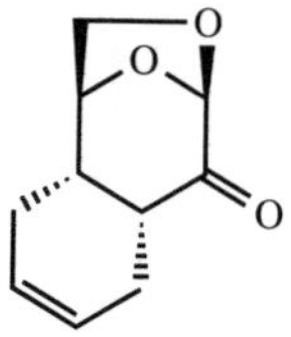

**6.41**
[79849-65-3][38]
Cellulose(2)

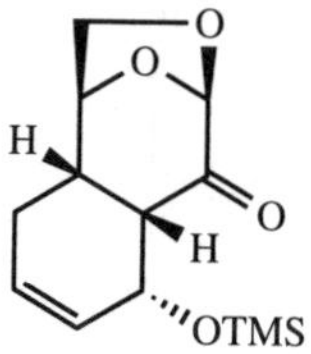

**6.42**
[—][39]
Cellulose(2)

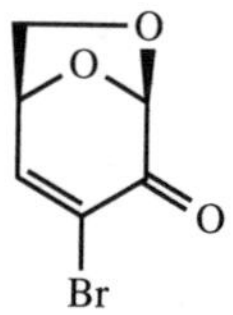

**6.43**
[78910-42-6][40]
Cellulose(2)

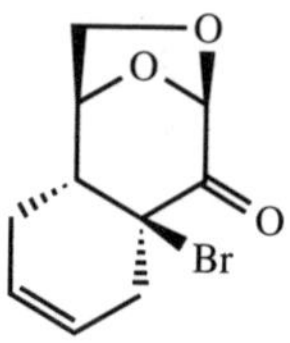

**6.44**
[79849-66-4][38]
Cellulose(2)

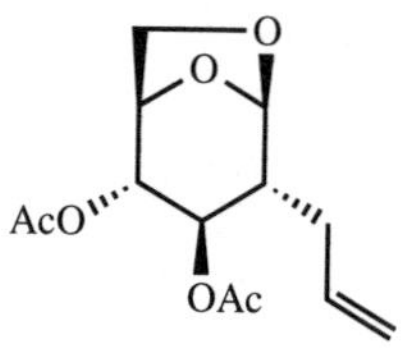

**6.45**
[149300-18-5][41]
D-Glucose(6)

## BUILDING BLOCK **6.46**

HO OH $N_3$

Name: β-D-Glucopyranose, 1,6-Anhydro-2-azido-2-deoxy-[67546-20-7]

### Synthesis:

D-Glucose

(a) $Ac_2O$, P, $Br_2$, 80-85%

(b) Zn, HOAc, 75-80%

(c) NaOMe, 73%

(d) $(Bu_3Sn)_2O$, $NaN_3$[26]

### Related Structures:

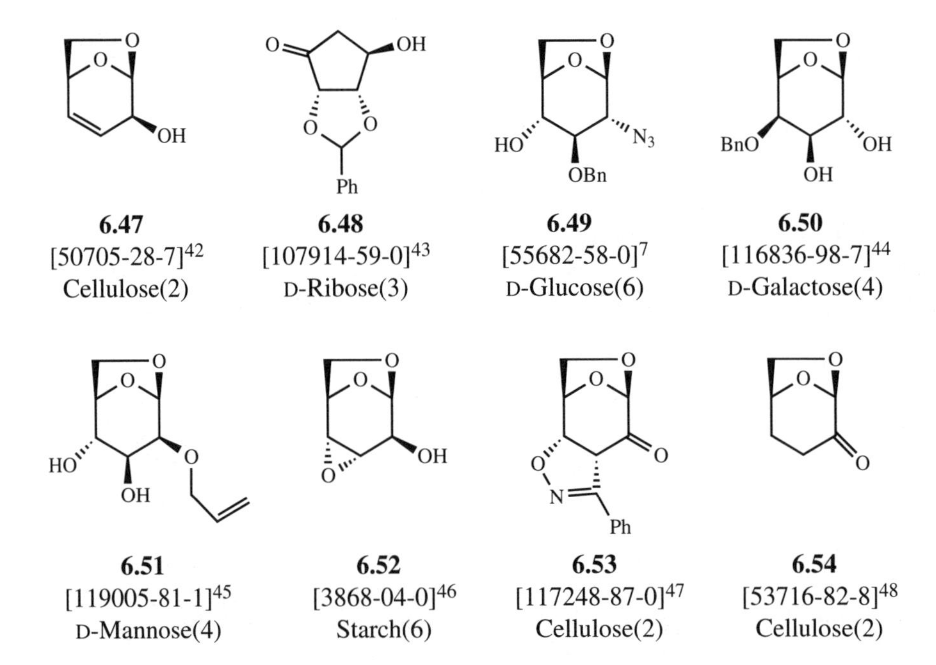

# BUILDING BLOCK 6.55

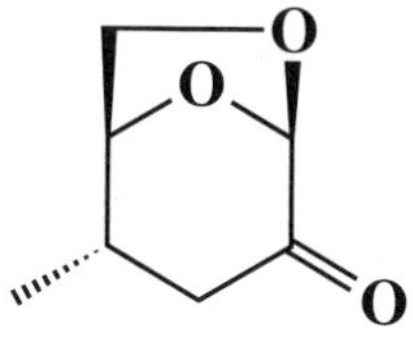

Name: β-D-*erythro*-Hexopyranos-2-ulose, 1,6-Anhydro-3,4-dideoxy-4-methyl-
[63000-65-7]

## Synthesis:

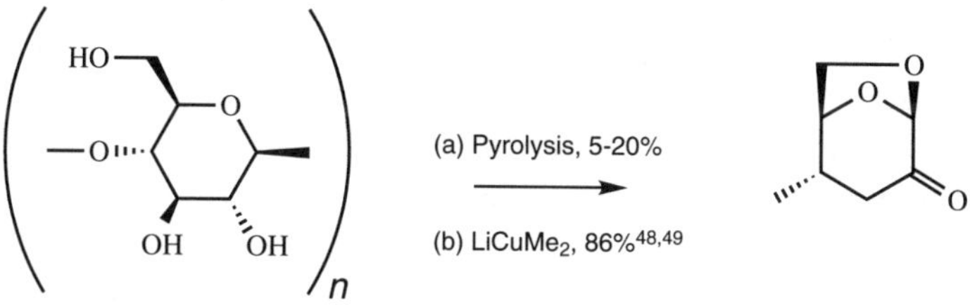

Cellulose

(a) Pyrolysis, 5-20%

(b) $LiCuMe_2$, 86%[48,49]

## Related Structures:

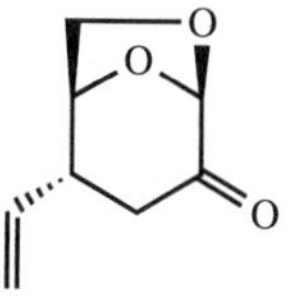

**6.56**
[—][50]
Cellulose(2)

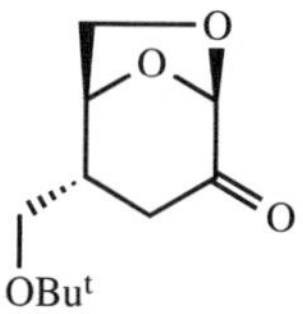

**6.57**
[—][51]
Cellulose(2)

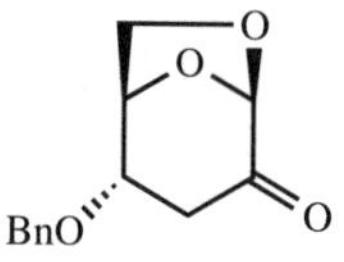

**6.58**
[—][51]
Cellulose(2)

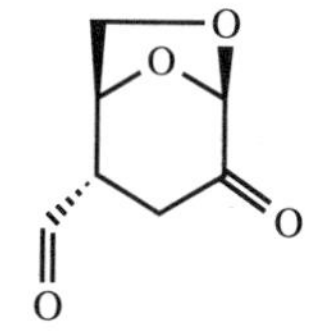

**6.59**
[—][50]
Cellulose(3)

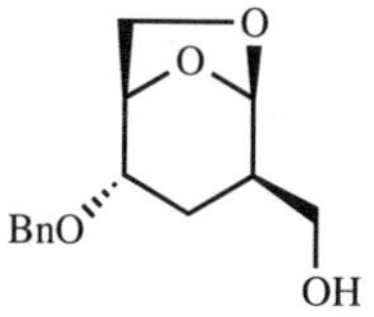

**6.60**
[—][51]
Cellulose(4)

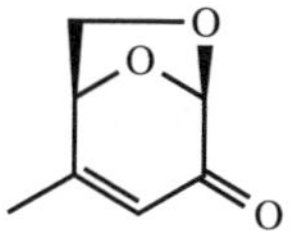

**6.61**
[151982-14-8][52]
Cellulose(4)

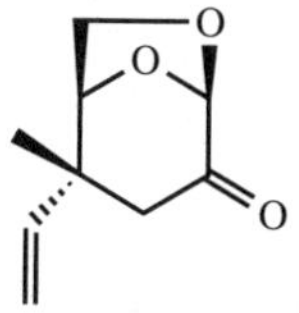

**6.62**
[151982-15-9][52]
Cellulose(5)

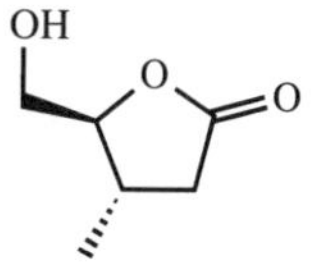

**6.63**
[81474-44-4][53]
Cellulose(3)

## REFERENCES


1. R. B. Ward. *Methods Carbohydr. Chem.* **1963** *2:*394–396.
2. M. Cerny and J. Stanek, Jr. *Adv. Carbohydr. Chem. Biochem.* **1974** *34:*23–177.
3. P. A. Gent, R. Gigg, and A. A. E. Penglis. *J. Chem. Soc. Perkin I* **1976** 1395–1404.
4. K. Heyns, P. Köll, and H. Paulsen. *Chem. Ber.* **1971** *104:*830–836.
5. A. E. Knauf, R. M. Hann, and C. S. Hudson. *J. Am. Chem. Soc.* **1941** *63:*1447–1451.
6. R. M. Hann and C. S. Hudson. *J. Am. Chem. Soc.* **1942** *64:*2435–2438.
7. H. Hori, Y. Nishida, H. Ohrui, and H. Meguro. *J. Org. Chem.* **1989** *54:*1346–1353.
8. K. Heyns, J. Weyer, and H. Paulsen. *Chem. Ber.* **1967** *100:*2317–2334.
9. B. Lindberg. *Methods Carbohydr. Chem.* **1972** *6:*323–325.
10. M. Cerny, L. Kalvoda, and J. Pacak. *Collect. Czech. Chem. Commun.* **1968** *33:*1143–1156.
11. M. Cerny, V. Gut, and J. Pacak. *Collect. Czech. Chem. Commun.* **1961** *26:*2542–2550.
12. T. Trnka and M. Cerny. *Collect. Czech. Chem. Commun.* **1971** *36:*2216–2225.
13. A. G. Kelly and J. S. Roberts. *J. Chem. Soc. Chem. Commun.* **1980** 228–229.
14. T. Inghardt and T. Frejd. *Synthesis* **1990** 285–291.
15. T. M. Jespersen, M. Bols, M. Sierks, and T. Skrydstrup. *Tetrahedron* **1994** *50:*13449–13460.
16. G. O. Aspinall and G. Zweifel. *J. Chem. Soc.* **1957** 2271–2278.
17. J. Stanek, Jr. and M. Cerny. *Synthesis* **1972** 698–699.
18. M. Cerny, T. Trnka, P. Beran, and J. Pacak. *Collect. Czech. Chem. Commun.* **1969** *34:*3377–3382.
19. H. B. Wood and H. G. Fletcher. *J. Am. Chem. Soc.* **1956** *78:*207–210.
20. R. J. Ferrier and R. H. Furneaux. *Aust. J. Chem.* **1980** *33:*1025–1036.
21. H. Ohrui, Y. Nishida, and H. Meguro. *Agric. Biol. Chem.* **1984** *48:*1049–1053.
22. H. Hori, T. Nakajima, Y. Nishida, H. Ohrui, and H. Meguro. *J. Carbohydr. Chem.* **1986** *5:*585–600.
23. G. H. Posner and S. R. Haines. *Tetrahedron Lett.* **1985** *26:*1823–1826.
24. H. Ohrui, T. Misawa, and H. Meguro. *Agric. Biol. Chem.* **1984** *48:*1825–1829.
25. K. Heyns and P. Köll. *Methods Carbohydr. Chem.* **1972** *6:*342–347.
26. D. Tailler, J.-C. Jacquinet, A.-M. Noirot, and J.-M. Beau. *J. Chem. Soc. Perkin I* **1992** 3163–3164.
27. F. Shafizadeh, R. H. Furneaux, and T. T. Stevenson. *Carbohydr. Res.* **1979** *71:*169–191.
28. R. H. Furneaux, J. M. Mason, and I. J. Miller. *J. Chem. Soc. Perkin I* **1984** 1923–1928.
29. E. L. Jackson and C. S. Hudson. *J. Am. Chem. Soc.* **1940** *62:*958–961.
30. M. Cerny and J. Pacak. *Collect. Czech. Chem. Commun.* **1962** *27:*94–105.
31. J. Pecka, J. Stanek, Jr., and M. Cerny. *Collect. Czech. Chem. Commun.* **1974** *39:*1192–1209.
32. P. A. Seib. *J. Chem. Soc. (C)* **1969** 2552–2559.

33. D. H. R. Barton and S.W. McCombie. *J. Chem. Soc. Perkin I* **1975** 1574–1585.
34. R. H. Bell, D. Horton, and D. M. Williams. *J. Chem. Soc. Chem. Commun.* **1968** 323–324.
35. T. Trnka and M. Cerny. *Collect. Czech. Chem. Commun.* **1972** *37:*3632–3639.
36. A. Klemer and M. Kohla. *Liebigs. Ann.* **1984** 1662–1671.
37. A. Klemer and M. Kohla. *Liebigs. Ann.* **1986** 967–979.
38. D. D. Ward and F. Shafizadeh. *Carbohydr. Res.* **1981** *95:*155–176.
39. M. Isobe, N. Fukami, T. Nishikawa, and T. Goto. *Heterocycles* **1987** *25:*521–532.
40. D. D. Ward and F. Shafizadeh. *Carbohydr. Res.* **1981** *93:*284–287.
41. C. Leteux, A. Veyrieres, and F. Robert. *Carbohydr. Res.* **1993** *242:*119–130.
42. K. Matsumoto, T. Ebata, K. Koseki, K. Okano, H. Kawakami, and H. Matsushita. *Carbohydr. Res.* **1993** *246:*345–352.
43. A. Klemer and M. Kohla. *Liebigs Ann.* **1987** 683–686.
44. M. C. Cruzado and M. Martin-Lomas. *Carbohydr. Res.* **1988** *175:*193–199.
45. F. Dasgupta and P. J. Garegg. *Synthesis* **1988** 626–628.
46. M. Cerny, J. Pacak, and J. Stanek. *Collect Czech. Chem. Commun.* **1965** *30:*1151–1157.
47. A. J. Blake, T. A. Cook, A. C. Forsyth, R. O. Gould, and R. M. Paton. *Tetrahedron* **1992** *48:*8053–8064.
48. F. Shafizadeh and P. P. S. Chin. *Carbohydr. Res.* **1977** *58:*79–87.
49. K. Mori, T. Chuman, and K. Kato. *Carbohydr. Res.* **1984** *192:*73–86.
50. Z. J. Witczak. In *Levoglucosenone and Levoglucosans, Chemistry and Applications,* Z. J. Witczak, ed., ATL Press, Mount Prospect, IL, **1994**, pp. 3–16.
51. R. Blattner, R. H. Furneaux, J. M. Mason, and P. C. Tyler. In *Levoglucosenone and Levoglucosans, Chemistry and Applications,* Z. J. Witczak, ed., ATL Press, Mount Prospect, IL, **1994**, pp. 43–57.
52. K. Okano, T. Ebata, K. Koseki, H. Kawakami, K. Matsumoto, and H. Matsushita. *Chem. Pharm. Bull.* **1993** *41:*861–865.
53. T. Ebata, K. Matsumoto, H. Yoshikoshi, K. Koseki, H. Kawakami, and H. Matsushita. *Heterocycles* **1990** *31:*1585–1588.

## BUILDING BLOCK **7.01**

Name: D-*arabino*-Hex-1-enitol, 1,5-Anhydro-2-deoxy-[13265-84-4]

### Synthesis:

D-Glucose

(a) $Br_2$, P, $Ac_2O$, $H^+$ 80-85%[1]

(b)Zn, HOAc, 75-80%[2]

(c) NaOMe, 73%[2]

### Related Structures:

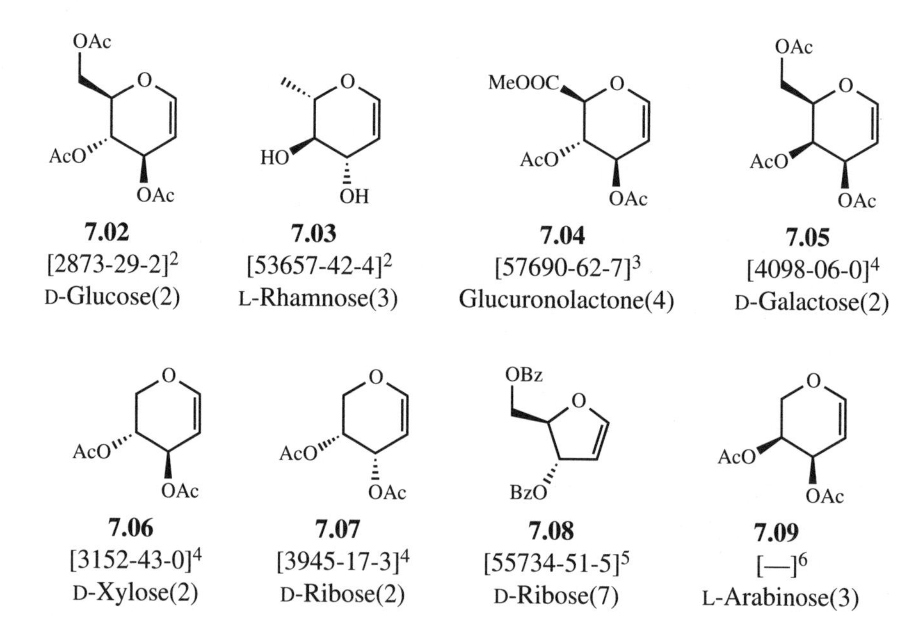

# BUILDING BLOCK 7.10

Name: D-*arabino*-Hex-1-enitol, 1,5-Anhydro-tetrabenzoate [114125-75-8]

## Synthesis:

D-Glucose

(a) BzCl, pyridine 95%[7]

(b) HBr-HOAc, 90%[8]

(c) NaI, $Et_2NH$ Acetone, 90%[9]

## Related Structures:

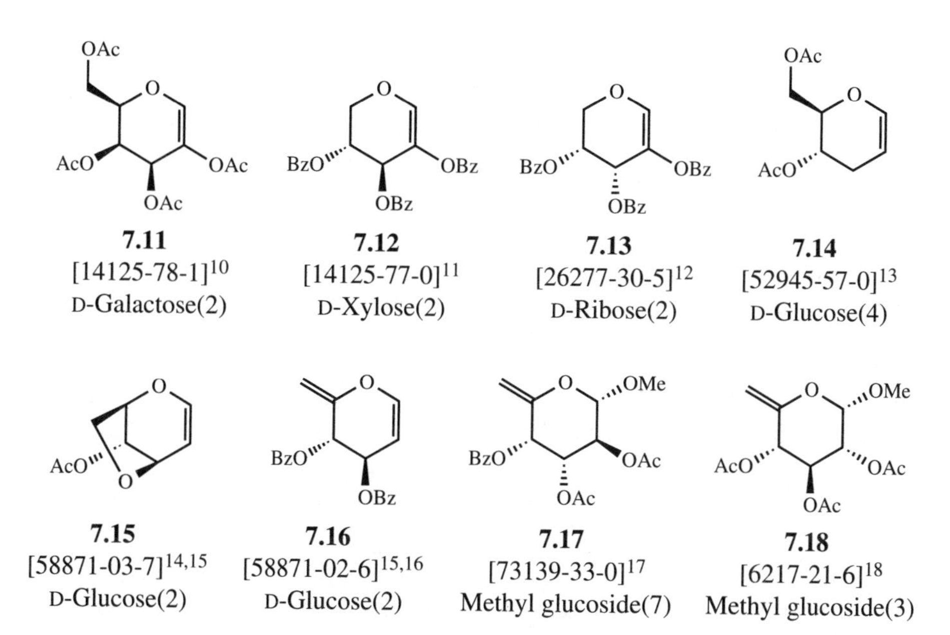

## BUILDING BLOCK **7.19**

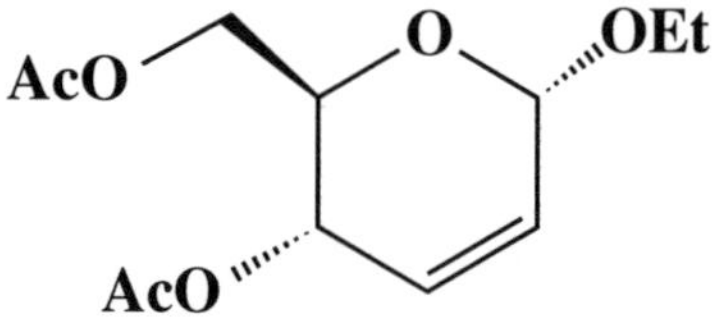

Name: α-D-*erythro*-Hex-2-enopyranoside, Ethyl 2,3-dideoxy-diacetate
[3323-72-6]

### Synthesis:

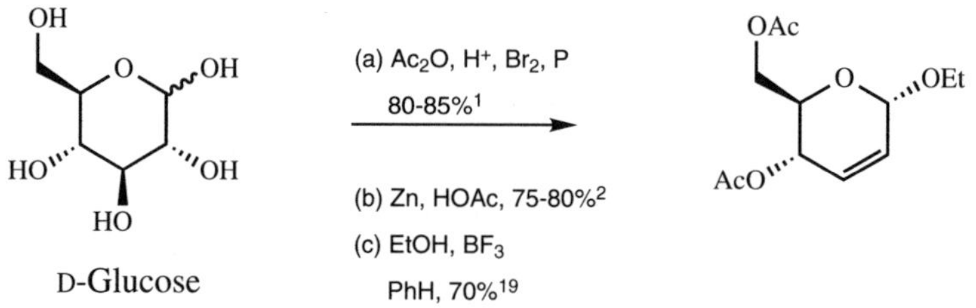

D-Glucose

(a) $Ac_2O$, $H^+$, $Br_2$, P
80-85%[1]

(b) Zn, HOAc, 75-80%[2]
(c) EtOH, $BF_3$
PhH, 70%[19]

### Related Structures:

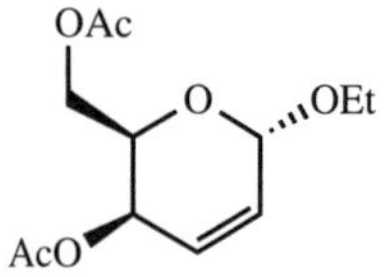

**7.20**
[69055-68-1][11]
D-Galactose(3)

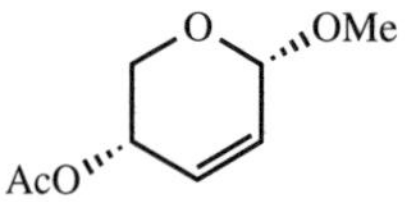

**7.21**
[25874-23-1][20]
D-Xylose(3)

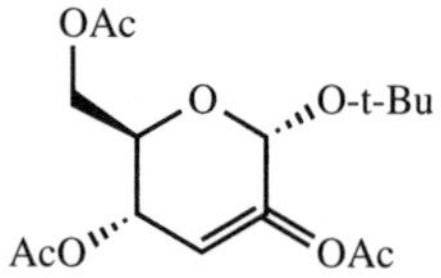

**7.22**
[79698-71-8][21]
D-Glucose(3)

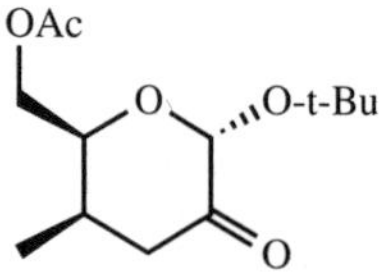

**7.23**
[81668-94-2][21]
D-Glucose(4)

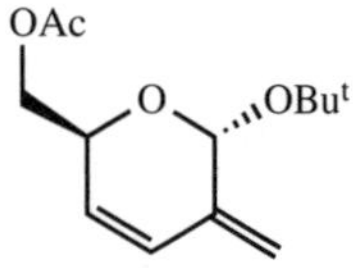

**7.24**
[79698-72-9][22]
D-Glucose(4)

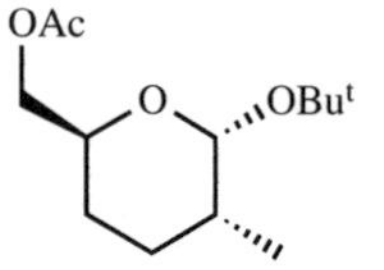

**7.25**
[79698-73-0][22]
D-Glucose(5)

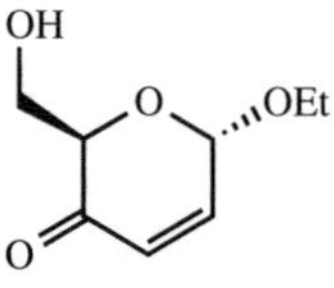

**7.26**
[25474-14-0][23]
D-Glucose(5)

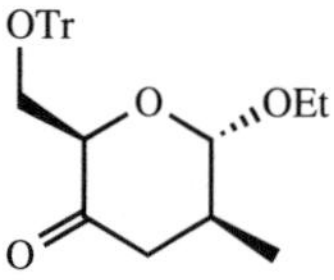

**7.27**
[66149-54-0][24]
D-Glucose(7)

# BUILDING BLOCK **7.28**

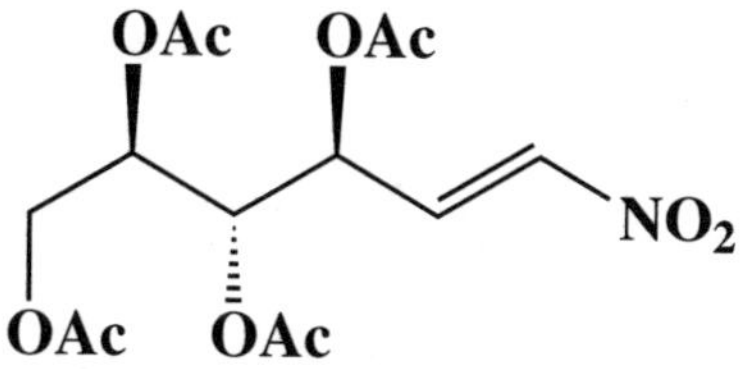
OAc
OAc
NO2
OAc
OAc

Name: D-*ribo*-Hex-2-enitol, 1,2-Dideoxy-1-nitro-3,4,5,6-tetraacetate, (1*E*)-[60478-51-5]

## Synthesis:

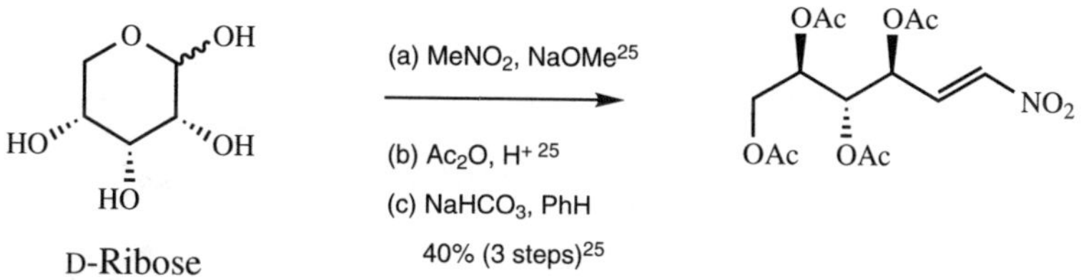
O
OH
HO
OH
HO
(a) MeNO2, NaOMe25
(b) Ac2O, H+ 25
(c) NaHCO3, PhH
40% (3 steps)25
OAc OAc
NO2
OAc OAc

D-Ribose

## Related Structures:

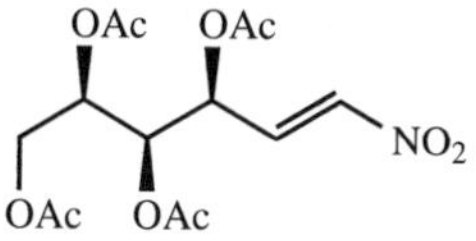
OAc OAc
NO2
OAc OAc

**7.29**
[39848-12-9][26]
D-Xylose(3)

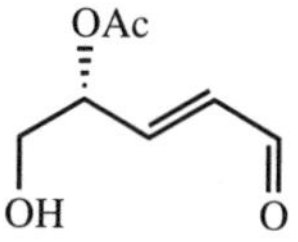
OAc
OH
O

**7.30**
[58886-30-9][27]
L-Arabinose(4)

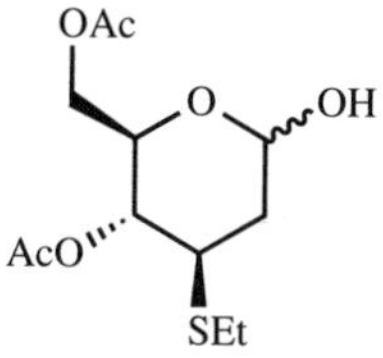
OAc
O
OH
AcO
SEt

**7.31**
[131853-24-2][28]
D-Glucose(4)

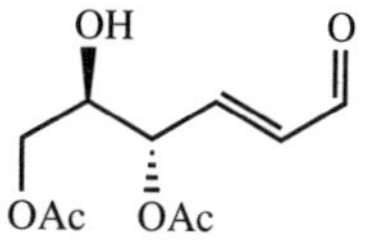
OH
O
OAc OAc

**7.32**
[29581-05-3][29]
D-Glucose(3)

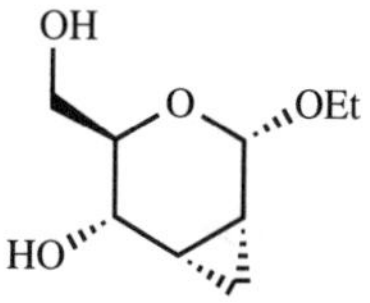
OH
O
OEt
HO

**7.33**
[39937-79-6][30]
D-Glucose(4)

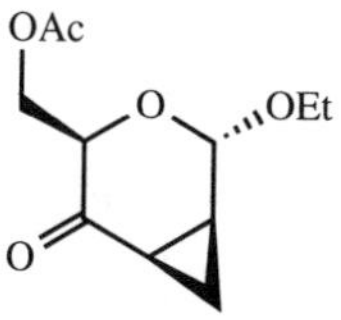
OAc
O
OEt
O

**7.34**
[39937-86-5][30]
D-Glucose(7)

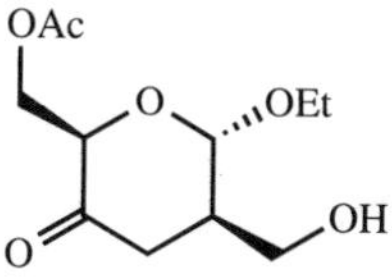
OAc
O
OEt
OH
O

**7.35**
[55941-69-0][31]
D-Glucose(7)

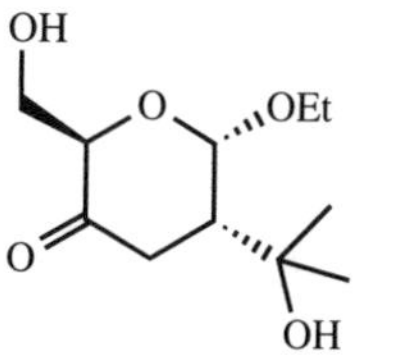
OH
O
OEt
O
OH

**7.36**
[40555-10-0][31]
D-Glucose(7)

## BUILDING BLOCK **7.37**

Name: D-*arabino*-Hex-1-enitol, 2,6-Anhydro tetrabenzoate [—]

### Synthesis:

D-Fructose

(a) BzCl, pyridine, 78%[32]

(b) HBr, $CH_2Cl_2$, 91%[32]

(c) DBU, MeCN, 81%[32]

### Related Structures:

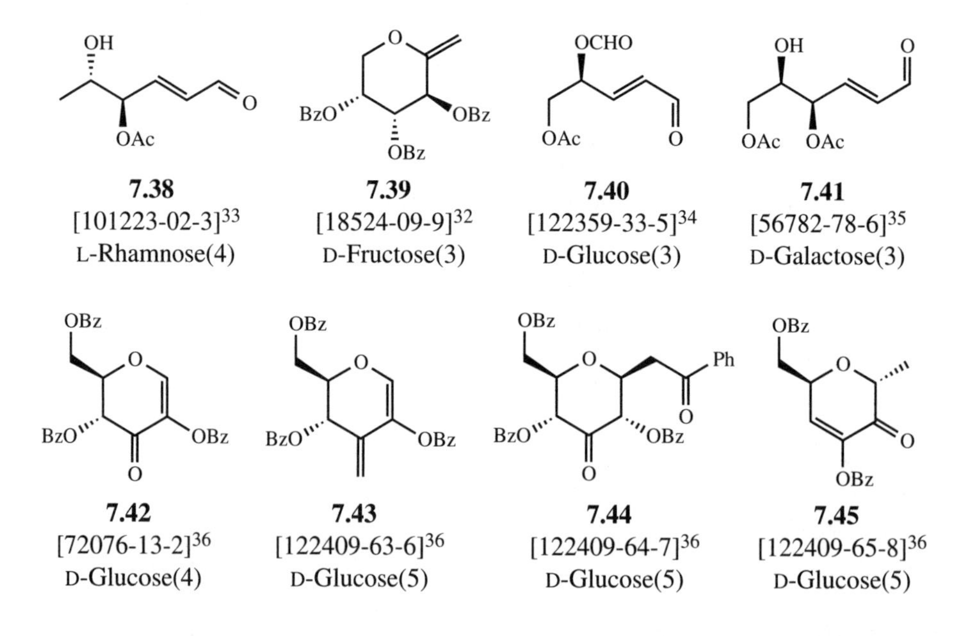

## BUILDING BLOCK **7.46**

OAc
O
O
AcO

Name: D-*erythro*-Hex-2-enonic acid, 2,3-Dideoxy-δ-lactone diacetate
[41976-28-7]

### Synthesis:

D-Glucose

(a) $Ac_2O$, $H^+$, $Br_2$, P
80-85%[1]

(b) Zn, HOAc, 75-80%[2]

(c) MCPBA, $BF_3$, 84%[34]

### Related Structures:

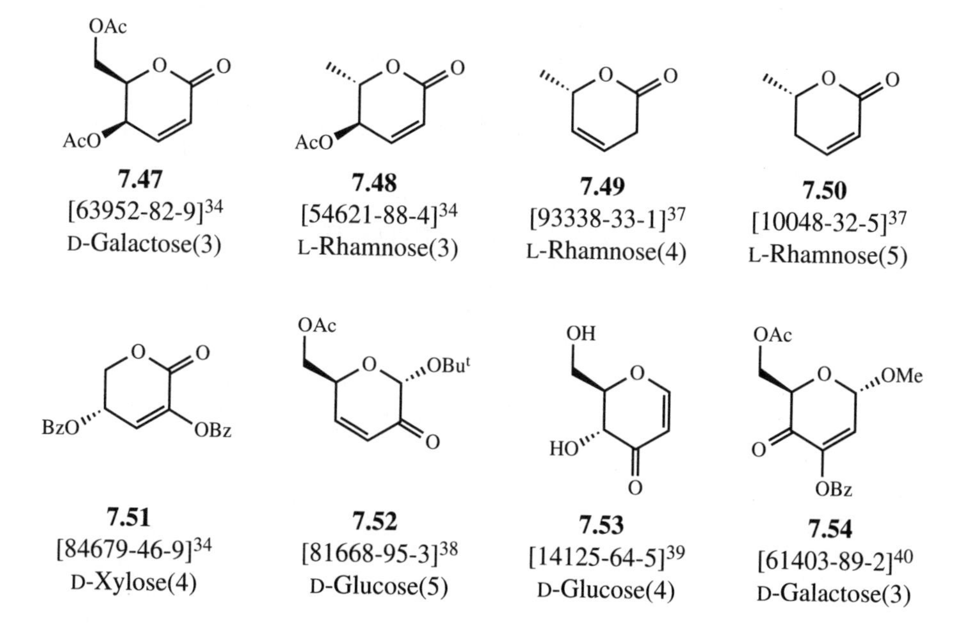

## BUILDING BLOCK **7.55**

Name: D-Arabinitol, 5-*C*-[5-(Acetyloxy)-6-nitro-3-cyclohexen-1-yl]-1,2,3,4,5-pentaacetate, [1*S*-(1$\alpha$(*R**),5$\beta$,5$\beta$)] [142561-72-6]

### Synthesis:

D-Galactose

(a) $MeNO_2$, NaOMe
(b) $Ac_2O$, $H^+$
(c) $NaHCO_3$,PhH
(d) $CH_2{=}CH{-}CH{=}CHOAc$[41]

### Related Structures:

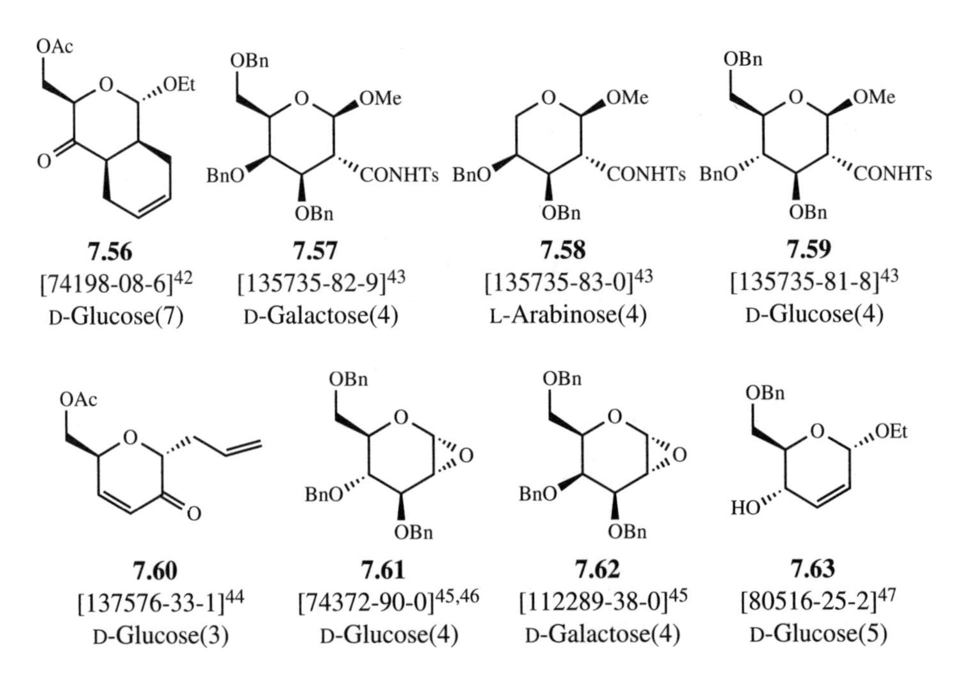

**7.56** [74198-08-6][42] D-Glucose(7)

**7.57** [135735-82-9][43] D-Galactose(4)

**7.58** [135735-83-0][43] L-Arabinose(4)

**7.59** [135735-81-8][43] D-Glucose(4)

**7.60** [137576-33-1][44] D-Glucose(3)

**7.61** [74372-90-0][45,46] D-Glucose(4)

**7.62** [112289-38-0][45] D-Galactose(4)

**7.63** [80516-25-2][47] D-Glucose(5)

## BUILDING BLOCK **7.64**

Name: Furane, 2,5-Dihydro-2-methoxy-5-[(triphenyl-methoxy)methyl]-
(2*R*-*cis*/2*S*-*trans*) [16802-02-1]

### Synthesis:

D-Xylose

(a) MeOH, $H^+$
(b) TrCl, pyridine
(c) S=C(imidazole)$_2$
(d) Raney Ni[48]

### Related Structures:

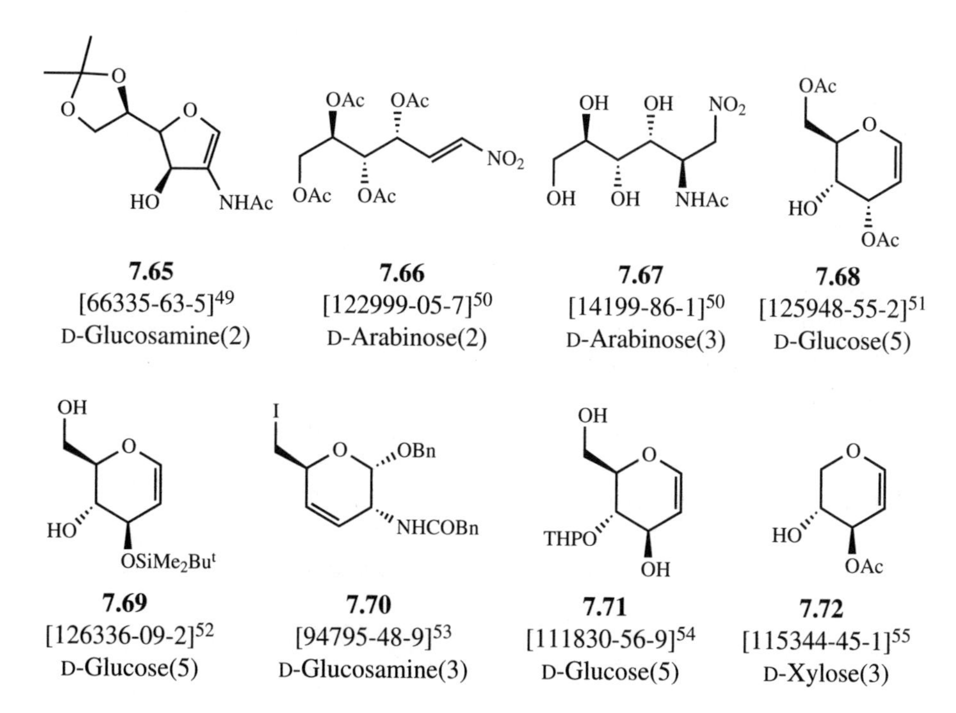

## REFERENCES


1. R. U. Lemieux. *Methods Carbohydr. Chem.* **1963** *2:*221–222.
2. W. Roth and W. Pigman. *Methods Carbohydr. Chem.* **1963** *2:*405–408.
3. H.-W. Fehlhaber, G. Snatzke, and I. Vlahov. *Liebigs Ann. Chim.* **1987** 637–638.
4. R. Csuk, A. Fürstner, B. I. Gläntzer, and H. Weidmann. *J. Chem. Soc. Chem. Commun.* **1986** 1149–1150.
5. R. K. Ness and H. G. Fletcher, Jr. *J. Org. Chem.* **1969** *28:*435–437.
6. W. C. Austin and F. L. Humoller. *J. Am. Chem. Soc.* **1934** *56:*1152–1153.
7. K. Maurer and R. Böhme. *Chem. Ber.* **1936** *69:*1399–1410.
8. H. G. Fletcher, Jr. *Methods Carbohydr. Chem.* **1963** *2:*226–228.
9. M. G. Blair. *Methods Carbohydr. Chem.* **1963** *2:*411–414.
10. O. Varela, G. M. de Fina, and R. M. de Lederkremer. *Carbohydr. Res.* **1987** *167:*187–196.
11. R. J. Ferrier and G. H. Sankey. *J. Chem. Soc. (C)* **1966** 2339–2345.
12. R. D. Rao and L. M. Lerner. *Carbohydr. Res.* **1972** *22:*345–350.
13. B. Fraser-Reid, B. Radatus, and S. Y.-K. Tam. *Methods Carbohydr. Chem.* **1980** *8:*219–225.
14. J. Kiss. *Carbohydr. Res.* **1969** *11:*579–581.
15. I. D. Blackburne, A. I. R. Burfitt, P. F. Fredericks, R. D. Guthrie. “Synthetic methods for carbohydrates.” *ACS Symp. Ser.* **1976** *39:*116–132.
16. E. M. Bessel. A. B. Foster, J. H. Westwood, L. D. Hall, and R. N. Johnson. *Carbohydr. Res.* **1971** *19:*39–48.
17. A. S. Machado, A. Olesker, and G. Lukacs. *Carbohydr. Res.* **1985** *135:*231–239.
18. A. Köhn and R. R. Schmidt. *Liebigs Ann. Chim.* **1987** 1045–1054.
19. R. J. Ferrier and N. Prasad. *J. Chem. Soc. (C)* **1969** 570–575.
20. R. J. Ferrier and N. Vethaviyaser. *J. Chem. Soc. Chem. Commun.* **1970** 1385–1387.
21. S. Hanessian, P. C. Tyler, and Y. Chapleur. *Tetrahedron Lett.* **1981** *22:*4583–4586.
22. S. Hanessian, P. C. Tyler, G. Demailly, and Y. Chapleur. *J. Am. Chem. Soc.* **1981** *103:*6243–6246.
23. B. Fraser-Reid, A. McLean, E.W. Usherwood, and M. Yunker. *Can. J. Chem.* **1970** *48:*2877–2884.
24. M. Yunker, D. E. Plaumann, and B. Fraser-Reid. *Can. J. Chem.* **1977** *55:*4002–4009.
25. M. B. Perry and J. Furdova. *Methods Carbohydr. Chem.* **1976** *7:*25–28.
26. M. B. Perry. *Methods Carbohydr. Chem.* **1976** *7:*29–31.
27. J. Wengel, J. Lau, and E. B. Pedersen. *Synthesis* **1989** 829–832.
28. J. Lau and E. B. Pedersen. *Acta Chem. Scand.* **1990** *44:*1046–1049.
29. B. Fraser-Reid and B. Radatus. *J. Am. Chem. Soc.* **1970** *92:*5288–5290.
30. B. Fraser-Reid and B. J. Carthy. *Can. J. Chem.* **1972** *50:*2928–2934.
31. B. Fraser-Reid, N. L. Holder, and M. B. Yunker. *J. Chem. Soc. Chem. Commun.* **1972** 1286–1287.
32. F.W. Lichtenthaler. In *Modern Synthetic Methods.* VCHA, Basel, **1992** 273–276.
33. J. D. White, E. G. Nolen, Jr., and C. H. Miller. *J. Org. Chem.* **1986** *51:*1150–1152.

34. F.W. Lichtenthaler, S. Rönninger, and P. Jarglis *Liebigs Ann. Chem.* **1989** 1153–1161.
35. F. Gonzales, S. Lesage, and A. S. Perlin. *Carbohydr. Res.* **1975** *42:*267–274.
36. F.W. Lichtenthaler, S. Nishiyama, and T. Weiner. *Liebigs Ann. Chem.* **1989** 1163–1170.
37. F.W. Lichtenthaler, F. D. Klingler, and P. Jarglis. *Carbohydr. Res.* **1984** *132:*C1–C4.
38. S. Hanessian, A.-M. Faucher, and S. Leger. *Tetrahedron.* **1990** *46:*231–243.
39. S. Czernecki, K. Vijayakumaran, and G. Ville. *J. Org. Chem.* **1986** *51:*5472–5475.
40. F.W. Lichtenthaler, S. Ogawa, and P. Heidel. *Chem. Ber.* **1977** *110:*3324–3332.
41. J. A. Serrano, L. E. Caceres, and E. Roman. *J. Chem. Soc. Perkin Trans. I* **1992** 941–942.
42. J. L. Primeau, R. C. Anderson, and B. Fraser-Reid. *J. Chem. Soc. Chem. Commun.* **1980** 6–8.
43. D. Mostowicz, O. Zegrocka, and M. Chmielewski. *Carbohydr. Res.* **1991** *212:*283–288.
44. R. J. Ferrier and P. M. Petersen. *J. Chem. Soc. Perkin Trans. I* **1992** 2023–2028.
45. R. H. Halcomb and S. J. Danishefsky. *J. Am. Chem. Soc.* **1989** *111:*6661–6666.
46. E. Wu and Q. Wu. *Carbohydr. Res.* **1993** *250:*327–333.
47. V. Pedretti, J.-M. Mallet, and P. Sinay. *Carbohydr. Res.* **1993** *244:*247–257.
48. B. Lacourt-Gadras, M. Grignon-Dubois, and B. Rezzonico. *Carbohydr. Res.* **1992** *235:*281–288.
49. A. Hasegawa and M. Kiso. *Carbohydr. Res.* **1979** *74:*341–344.
50. J. C. Sowden and M. L. Oftedahl. *Methods Carbohydr. Chem.* **1962** *1:*235–237.
51. M. D. Wittman, R. H. Halcomb, S. J. Danishefsky, J. Gulik, and D. Vyas. *J. Org. Chem.* **1990** *55:*1979–1981.
52. M. S. Shelchani, K. M. Khan, K. Mahmood, P. M. Shah, and S. Malik. *Tetrahedron Lett.* **1990** *31:*1669–1670.
53. Z. Pakulski and A. Zamojski. *Carbohydr. Res.* **1990** *205:*410–414.
54. S. Bouhroum and P. J. A. Vottero. *Tetrahedron Lett.* **1987** *28:*5529–5530.
55. W. Kinzy and R.R. Schmidt. *Tetrahedron Lett.* **1987** *28:*1981–1984.

## BUILDING BLOCK **8.01**

Name: D-Ribonic acid, 2-*C*-Methyl $\gamma$-lactone [492-30-8]

### Synthesis:

$Ca(OH)_2$, 9%[1]

D-Fructose

### Related Structures:

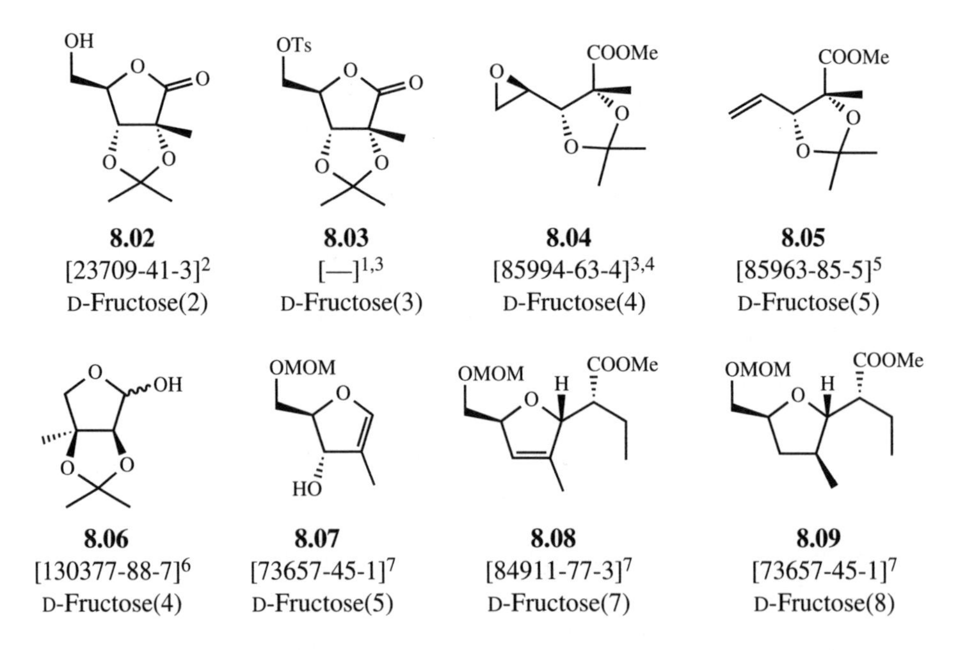

## BUILDING BLOCK **8.10**

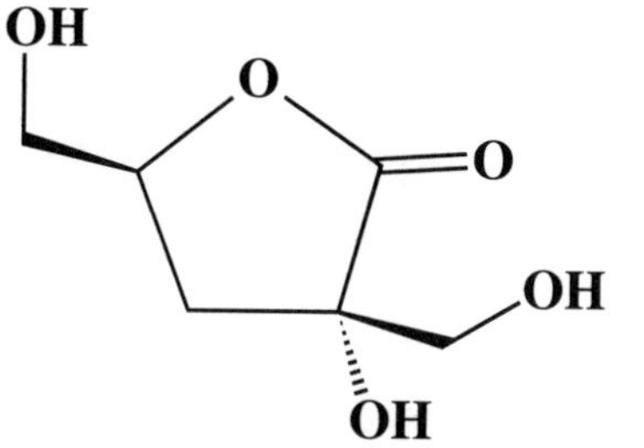

Name: D-*erythro*-Pentonic acid, 3-Deoxy-2-*C*-hydroxymethyl γ-lactone
[7397-89-9]

### Synthesis:

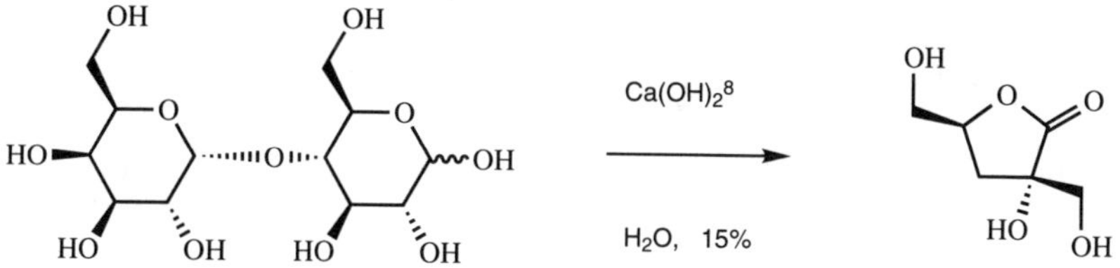

Lactose → $Ca(OH)_2$[8], $H_2O$, 15%

### Related Structures:

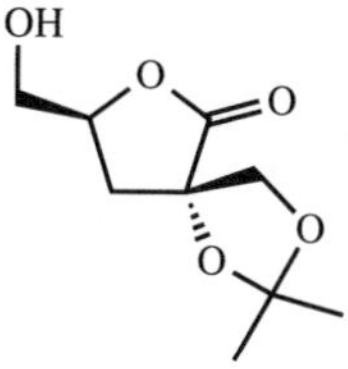

**8.11**
[78687-63-5][9]
Lactose(2)

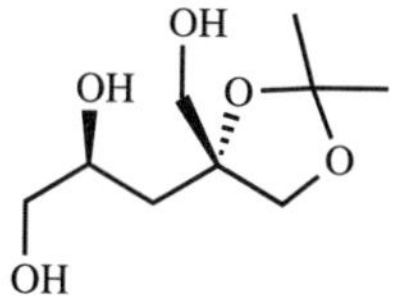

**8.12**
[93662-50-1][10]
Lactose(3)

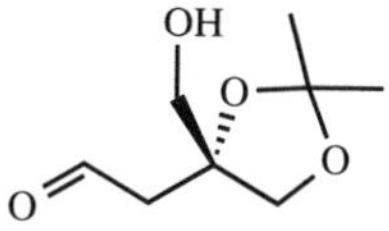

**8.13**
[95335-02-7][10]
Lactose(4)

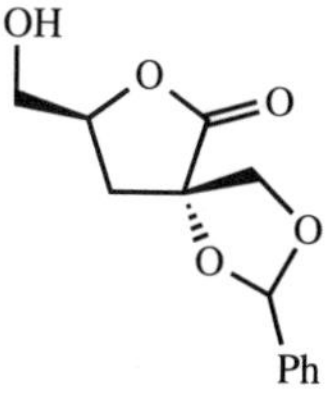

**8.14**
[—][11]
Lactose(2)

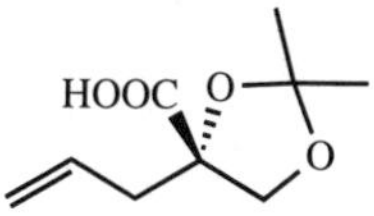

**8.15**
[106267-98-5][12]
Lactose(5)

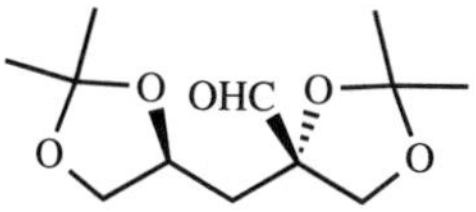

**8.16**
[93662-54-5][13]
Lactose(4)

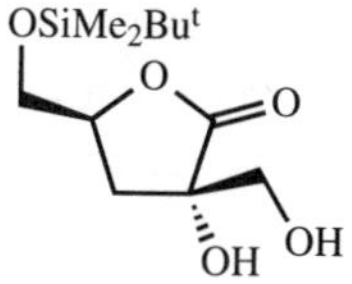

**8.17**
[—][14]
Lactose(2)

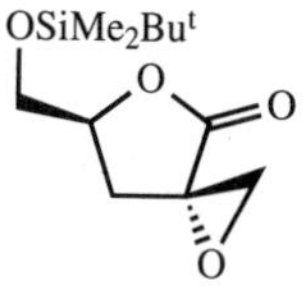

**8.18**
[—][14]
Lactose(4)

## BUILDING BLOCK **8.19**

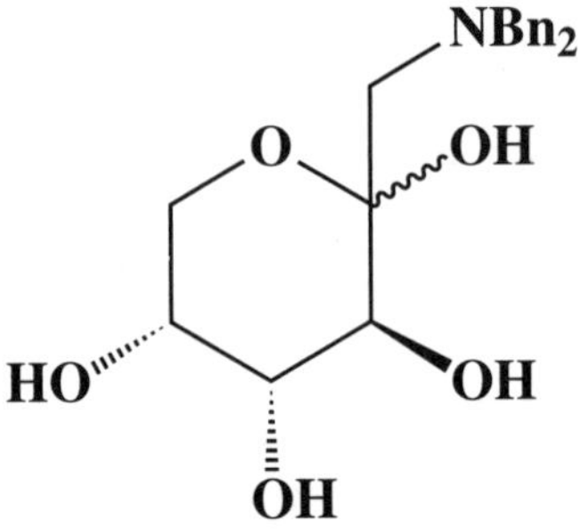
NBn2
O
OH
HO
OH
OH

Name: D-Fructose, 1-Deoxy-1-(dimethylphenylamino)-[69712-22-7]

### Synthesis:

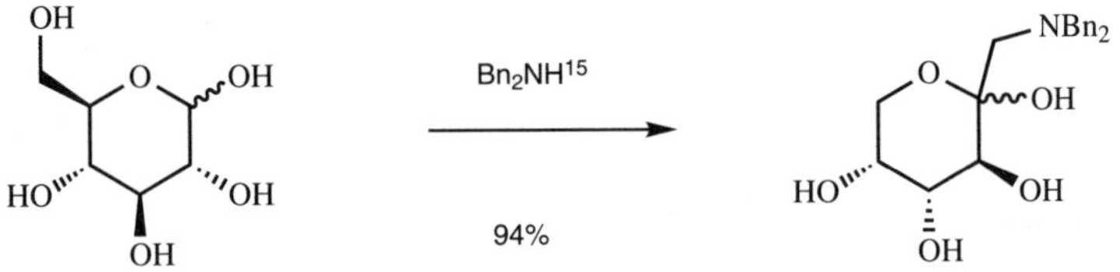
OH
O
OH
HO
OH
OH
Bn2NH15
94%
NBn2
O
OH
HO
OH
OH

D-Glucose

### Related Structures:

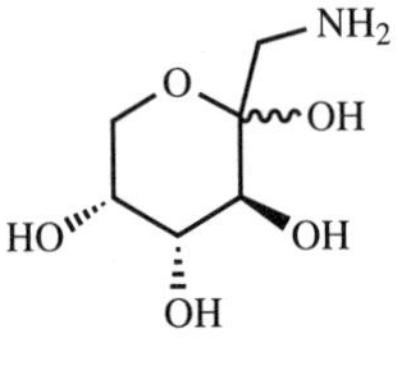

**8.20**
[4429-04-3][15]
Glucose(2)

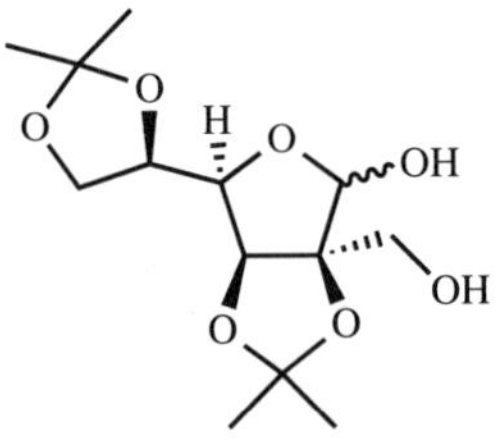

**8.21**
[91738-34-0][16]
Mannose(2)

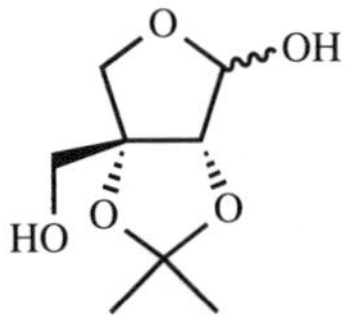

**8.22**
[94943-41-6][16]
Mannose(5)

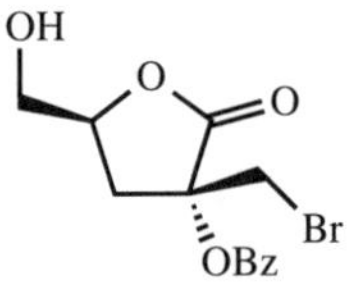

**8.23**
[—][11]
Lactose(3)

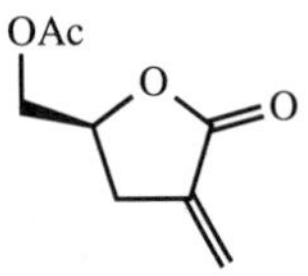

**8.24**
[111507-14-3][17]
Lactose(4)

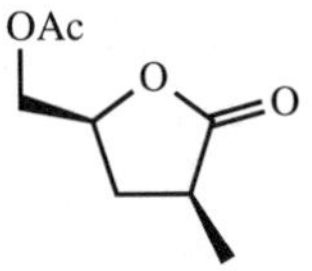

**8.25**
[111507-13-2][17]
Lactose(5)

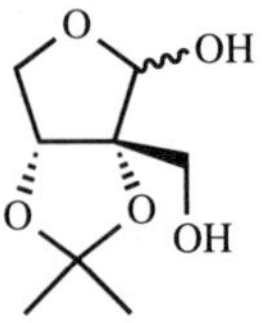

**8.26**
[—][18,19]
D-Arabinose(5)

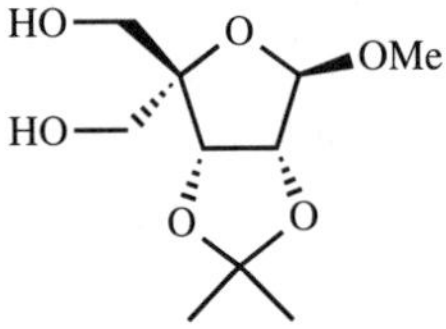

**8.27**
[55797-67-6][20]
D-Ribose(4)

# BUILDING BLOCK **8.28**

Name: L-Fructose, 6-Deoxy-[14807-05-7]

## Synthesis:

L-Rhamnose → (Pyridine[21], reflux, 95-97%)

## Related Structures:

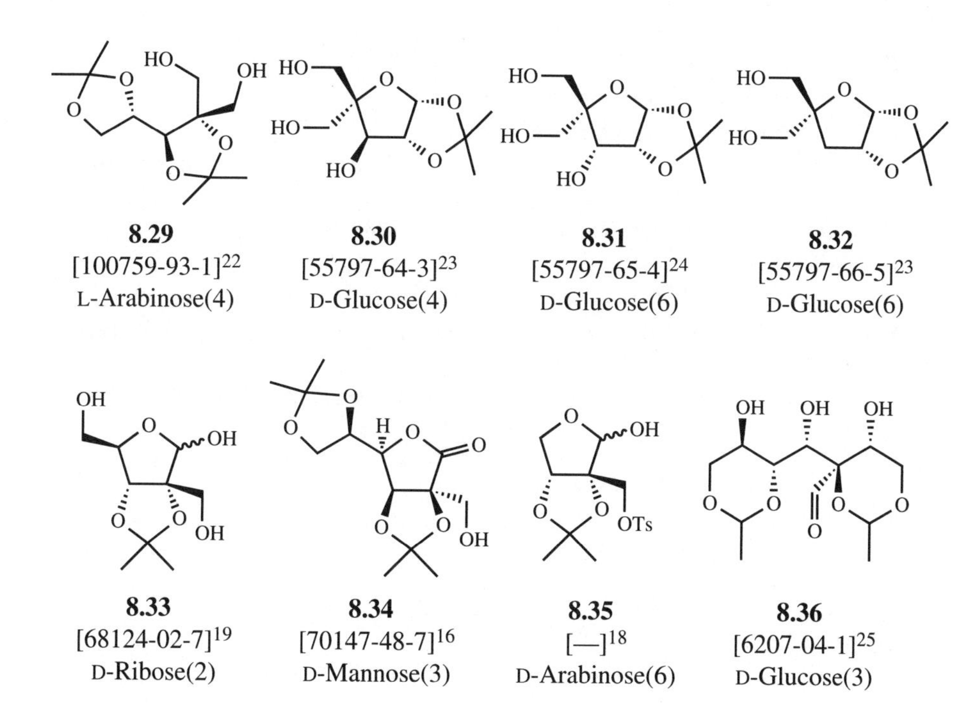

## REFERENCES

1. R. L. Whistler and J. N. BeMiller. *Methods Carbohydr. Chem.* **1963** *2:*484–485.
2. R. E. Ireland, R. C. Andersen, R. Baboud, B. J. Fitzsimmons, G. J. McGarvey, S. Thaisrivongs, and C. S. Wilcox. *J. Am. Chem. Soc.* **1983** *105:*1988–2006.
3. J. C. Sowden and R. Strobach. *J. Am. Chem. Soc.* **1960** *82:*3707–3709.
4. R.W. Hoffman and W. Ladner. *Chem. Ber.* **1983** *116:*1631–1642.
5. B. Deguin, J.-C. Florent, and C. Monneret. *J. Org. Chem.* **1991** *56:*405–411.
6. K. Ando, T. Yamada, Y. Takaishi, and M. Shibuya. *Heterocycles* **1989** *29:*1023–1027.
7. R. E. Ireland, R. C. Andersen, R. Baboud, B. J. Fitzsimmons, G. J. McGarvey, S. Thaisrivongs, and C. S. Wilcox. *J. Am. Chem. Soc.* **1983** *105:*1988–2006.
8. R. L. Whistler and J. N. BeMiller. *Methods Carbohydr. Chem.* **1963** *2:*477–479.
9. S. Hanessian and R. Roy. *Tetrahedron Lett.* **1981** *22:*1005–1008.
10. F. Bennani, J.-C. Florent, M. Koch, and C. Monneret. *Tetrahedron* **1984** *40:*4669–4676.
11. C. Monneret and J.-C. Florent. *Synlett* **1994** 310.
12. J.-C. Florent, J. Ughetto-Monfrin, and C. Monneret. *J. Org. Chem.* **1987** *52:*1051–1056.
13. K. Matsuo, Y. Hasuike, and H. Kado. *Chem. Pharm. Bull.* **1990** *33:*2847–2849.
14. C. Monneret and J.-C. Florent. *Synlett* **1994** 305–318.
15. J. E. Hodge and B. E. Fisher. *Methods Carbohydr. Chem.* **1963** *2:*99–107.
16. P.-T. Ho. *Can. J. Chem.* **1979** *57:*381–383.
17. K. Bock, I. M. Castilla, I. Lundt, and C. Pedersen. *Acta Chem. Scand.* **1987** *B41:*13–17.
18. P.-T. Ho. *Can. J. Chem.* **1980** *58:*858–860.
19. P.-T. Ho. *Tetrahedron Lett.* **1978** 1623–1626.
20. G. H. Jones, M. Taniguchi, D. Tegg, and J. G. Moffatt. *J. Org. Chem.* **1979** *44:*1309–1317.
21. S. Ennifar and H. S. El Khadem. *Carbohydr. Res.* **1989** *193:*303–306.
22. D. T. Williams and J. K. N. Jones. *Can. J. Chem.* **1964** *42:*69–72.
23. D. L. Leland and M. P. Kotick. *Carbohydr. Res.* **1974** *38:*C9–C11.
24. R. D. Youssefyeh, J. P. H. Verheyden, and J. G. Moffatt. *J. Org. Chem.* **1979** *44:*1301–1309.
25. R. Schaffer and H. S. Isbell. *Methods Carbohydr. Chem.* **1962** *1:*273–275.

## BUILDING BLOCK **9.01**

HO
O
OH

Name: 1,2-Ethanediol, 1-(2-Furyl)-D-*glycero*-[14086-08-9]

### Synthesis:

OH O OH HO OH OH
D-Glucose

(a) $Ac_2O$, $H^+$, $Br_2$, P, 80-85%[1]
(b) Zn, AcOH, 75-82%[2]
(c) NaOMe, 73%[2]
(d) $H^+$, $HgSO_4$, 81%[3]

HO O OH

### Related Structures:

| **9.02** | **9.03** | **9.04** | **9.05** |
|---|---|---|---|
| [16299-15-3][4] | [29884-71-7][4] | [19186-39-1][4] | [120269-53-6][5] |
| Glucose(4) | Glucose(4) | Galactose(4) | Glucose(5) |

## REFERENCES


1. R. U. Lemieux. *Methods Carbohydr. Chem.* **1963** *2:*221–222.
2. W. Roth and W. Pigman. *Methods Carbohydr. Chem.* **1963** *2:*405–408.
3. F. Gonzales, S. Lesage, and A. S. Perlin. *Carbohydr. Res.* **1975** *42:*267–274.
4. H. Paulsen. *Methods Carbohydr. Chem.* **1972** *6:*142–148.
5. F. M. Hauser, S. R. Ellenberger, and W. P. Ellenberger. *Tetrahedron Lett.* **1988** *29:*4939–4942.

BUILDING BLOCK **10.01**

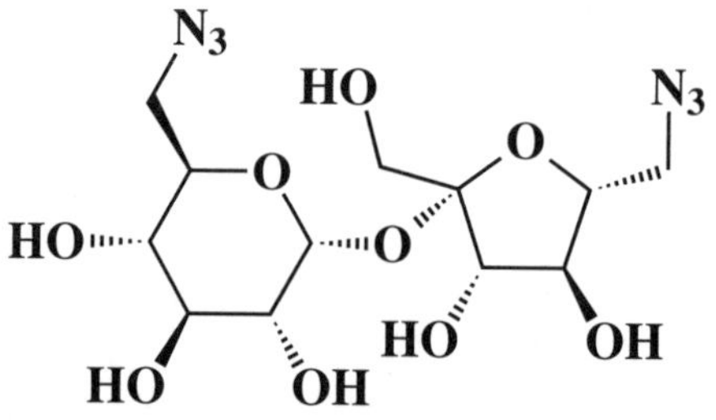

Name: α-D-Glucopyranoside, 6-Azido-6-deoxy-β-fructofuranosyl, 6-Azido-6-deoxy-[—]

**Synthesis:**

Sucrose

(a) $Ph_3P$, $CCl_4$ pyridine, 65-75 %[1]

(b) $NaN_3$, DMF, 81%[1]

**Related Structures:**

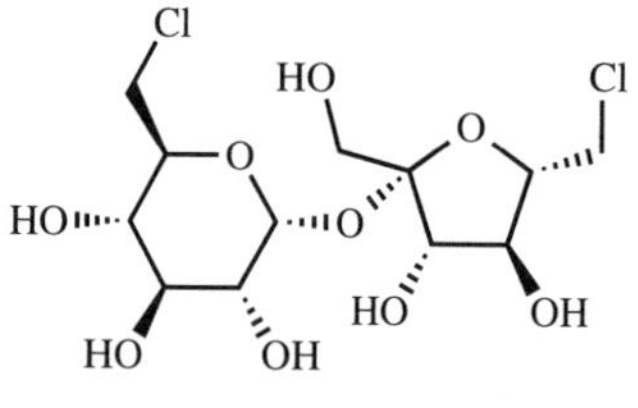

**10.02**
[40984-16-5][2]
Sucrose(1)

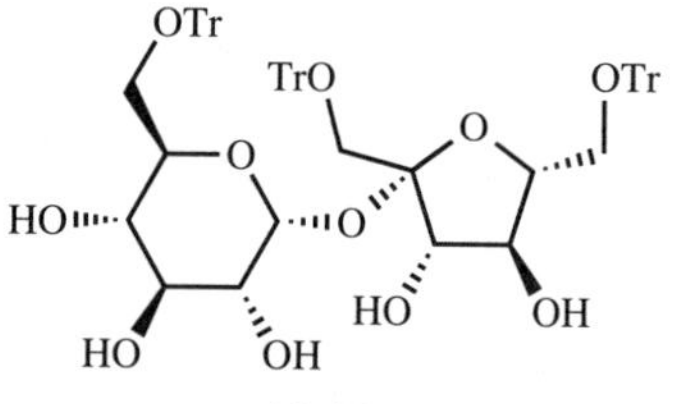

**10.03**
[35674-14-7][3]
Sucrose(1)

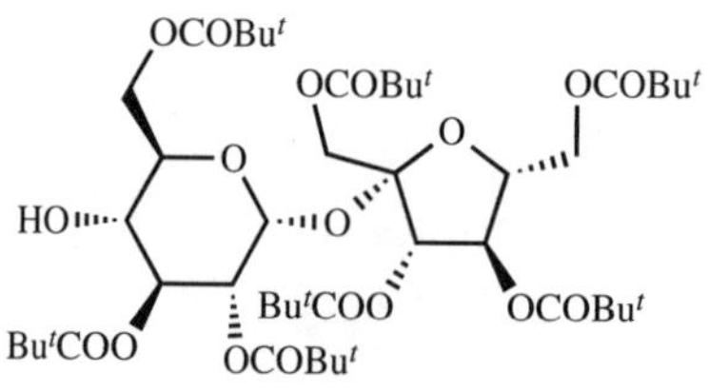

**10.04**
[69075-32-7][4]
Sucrose(1)

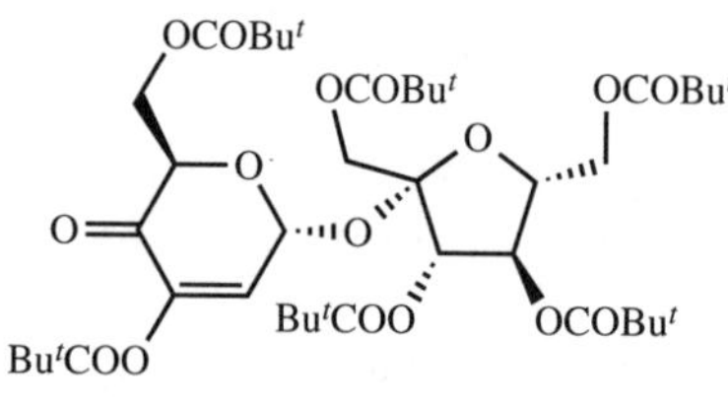

**10.05**
[—][5]
Sucrose(3)

## BUILDING BLOCK **10.06**

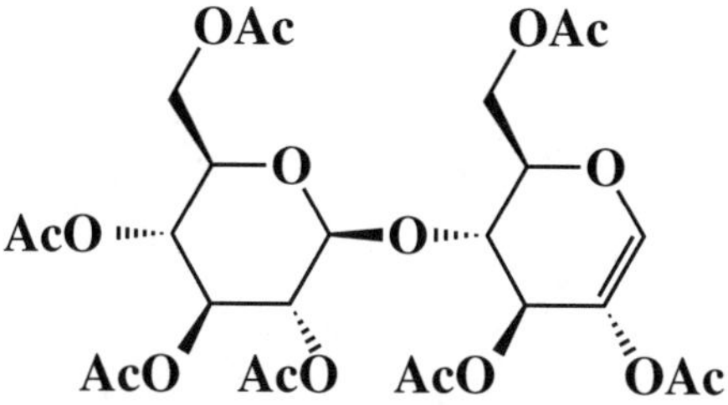

Name: D-*arabino*-Hex-1-enitol, 1,5-Anhydro-4-*O*-(2,3,4,6-tetra-*O*-acetyl-β-D-glucopyranosyl)triacetate [35526-17-1]

### Synthesis:

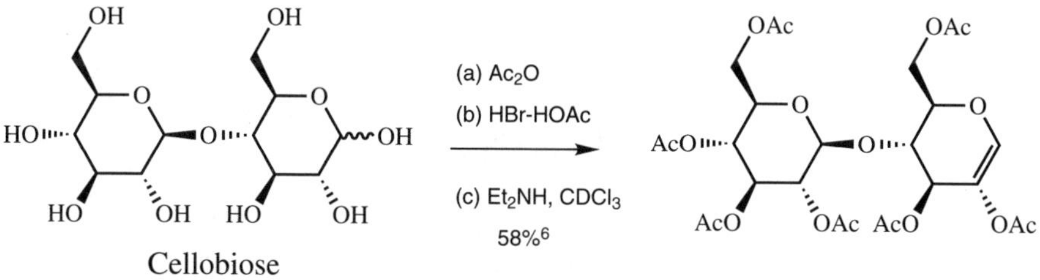

Cellobiose

(a) $Ac_2O$
(b) HBr-HOAc
(c) $Et_2NH$, $CDCl_3$
58%[6]

### Related Structures:

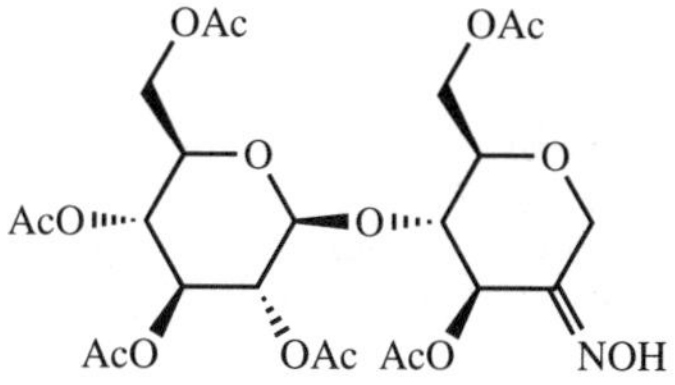

**10.07**
[—][5]
Cellobiose(4)

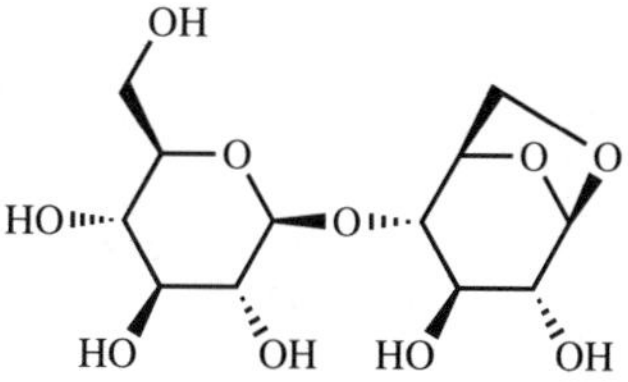

**10.08**
[35405-71-1][7]
Cellobiose(3)

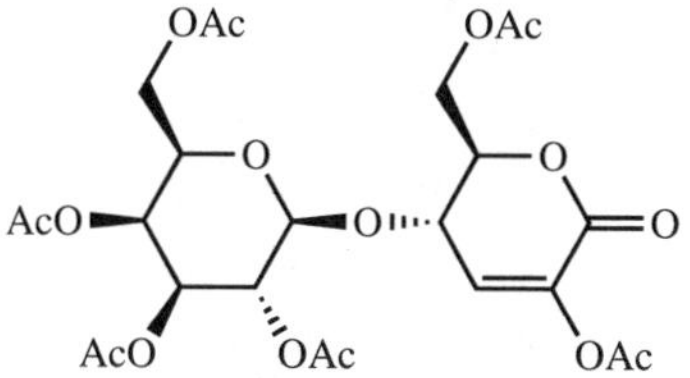

**10.09**
[122517-17-3][8]
Lactose(4)

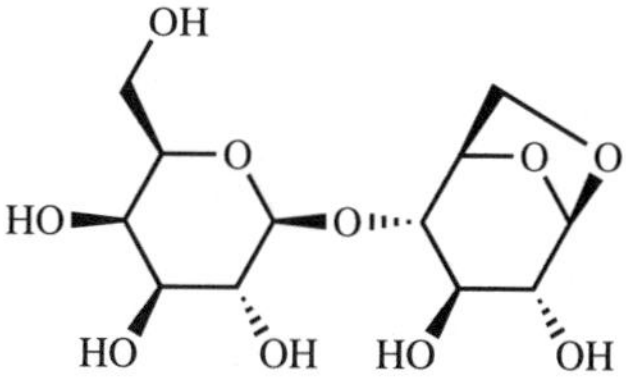

**10.10**
[34395-01-2][9]
Lactose(3)

## BUILDING BLOCK **10.11**

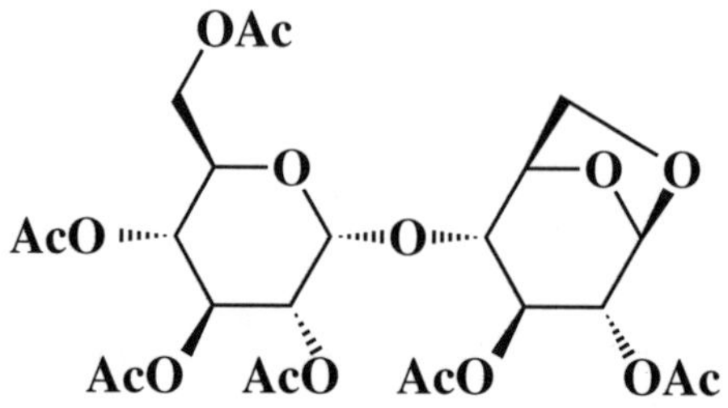
OAc
AcO
O
O
O
O
AcO
AcO
AcO
OAc

Name: β-D-Glucopyranose, 1,6-Anhydro-4-*O*-α-D-glucopyranosyl-hexaacetate
[28868-67-9]

### Synthesis:

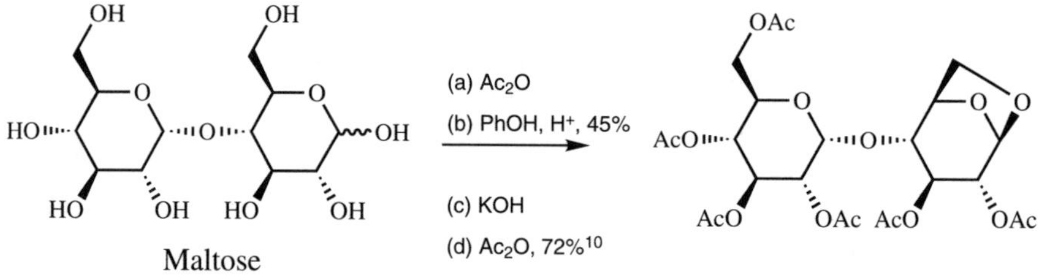
Maltose
(a) $Ac_2O$
(b) PhOH, $H^+$, 45%
(c) KOH
(d) $Ac_2O$, 72%[10]

### Related Structures:

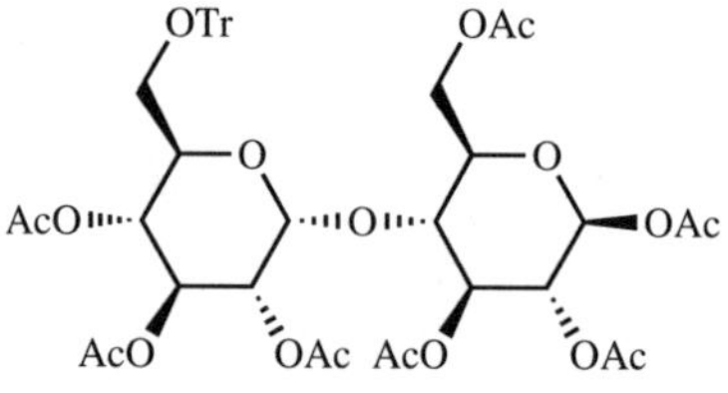

**10.12**
[6748-73-8][11]
Maltose(1)

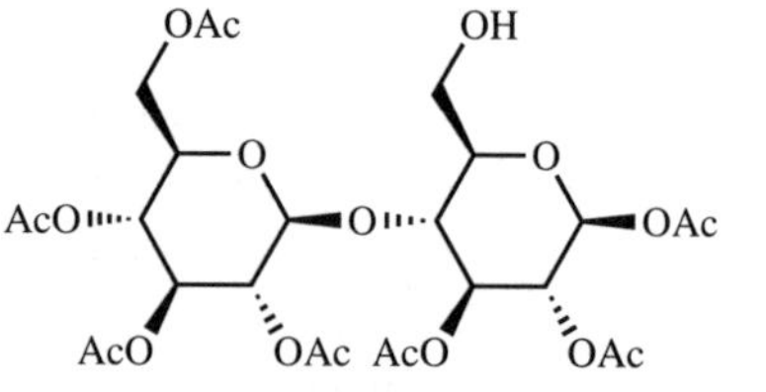

**10.13**
[7482-60-2][12]
Maltose(4)

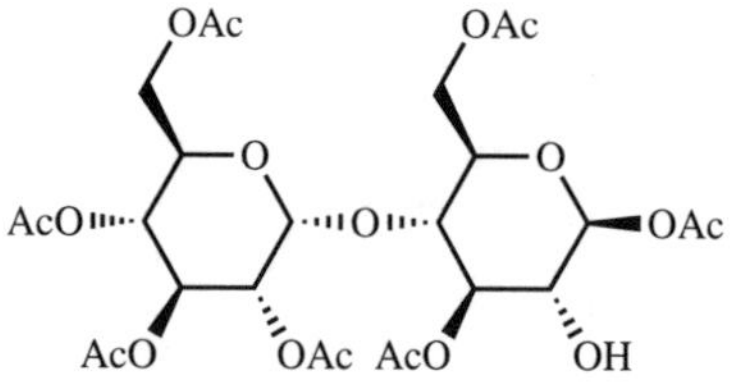

**10.14**
[32447-67-9][13]
Maltose(4)

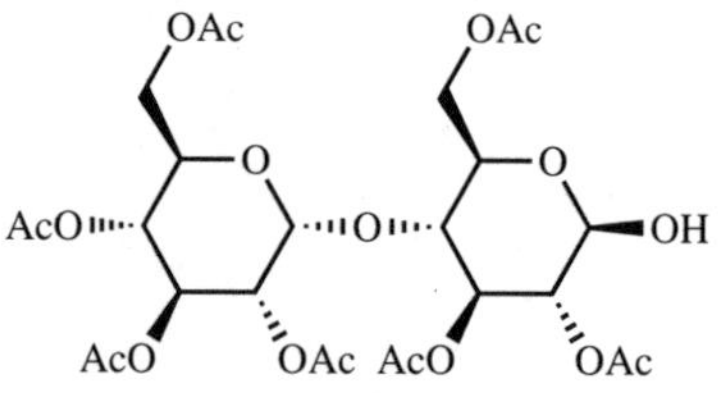

**10.15**
[18423-98-8][14]
Maltose(2)

## BUILDING BLOCK **10.16**

Name: β-Maltose, 1,2,2′,3′,4′,6,6′-Heptaacetate [19204-77-4]

**Synthesis:**

Maltose

$Ac_2O$[15]

71%

**Related Structures:**

**10.17**
[117399-13-0][16]
Maltose(4)

**10.18**
[72076-14-3][17]
Maltose(2)

**10.19**
[67314-34-5][18]
Maltose(4)

**10.20**
[56665-80-6][19]
Maltose(2)

## BUILDING BLOCK **10.21**

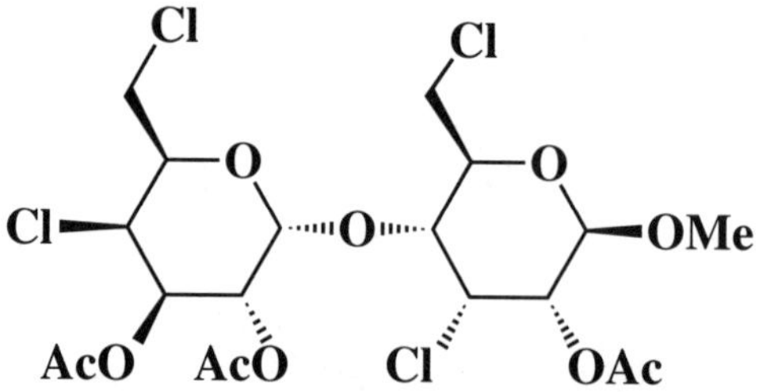

Cl Cl O Cl O O OMe AcO AcO Cl OAc

Name: β-D-Allopyranoside, Methyl 3,6-dichloro-3,6-dideoxy-4-*O*-(2,3-di-*O*-acetyl-4,6-dichloro-4,6-dideoxy-α-D-galactopyranosyl)acetate [51532-95-7]

### Synthesis:

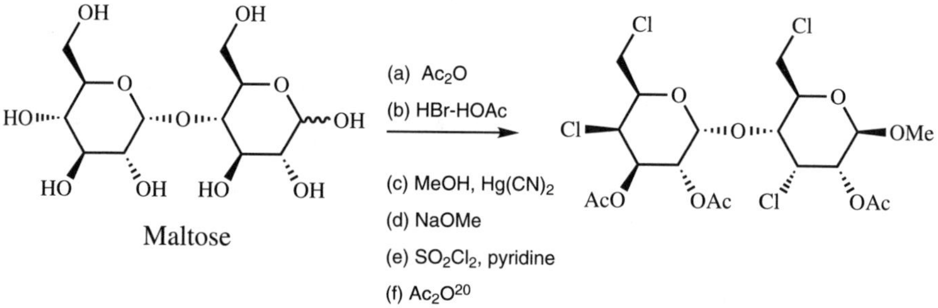

Maltose

(a) $Ac_2O$
(b) HBr-HOAc
(c) MeOH, $Hg(CN)_2$
(d) NaOMe
(e) $SO_2Cl_2$, pyridine
(f) $Ac_2O$[20]

### Related Structures:

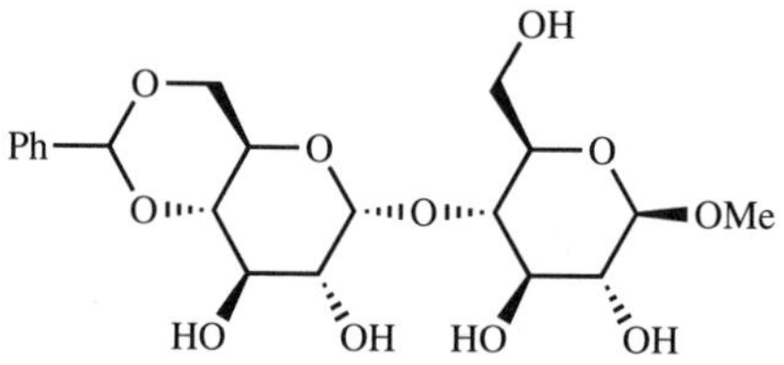

**10.22**
[57472-04-5][21]
Maltose(5)

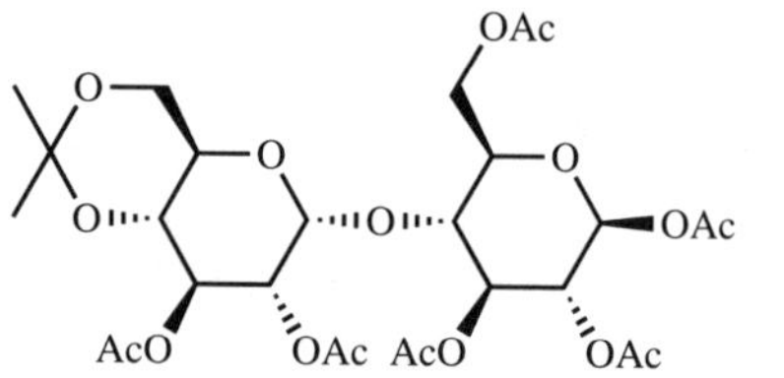

**10.23**
[74135-02-7][22]
Maltose(1)

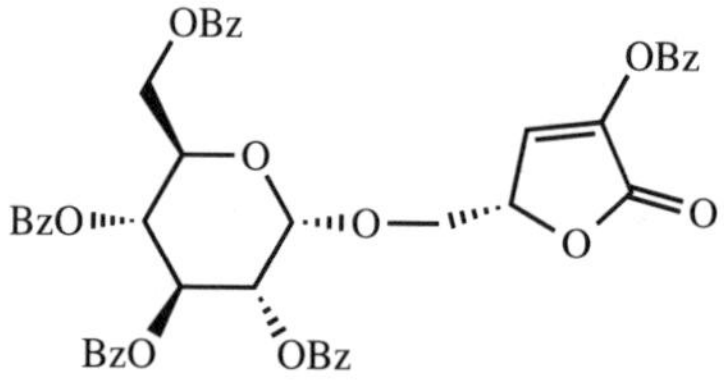

**10.24**
[—][5]
Isomaltulose(4)

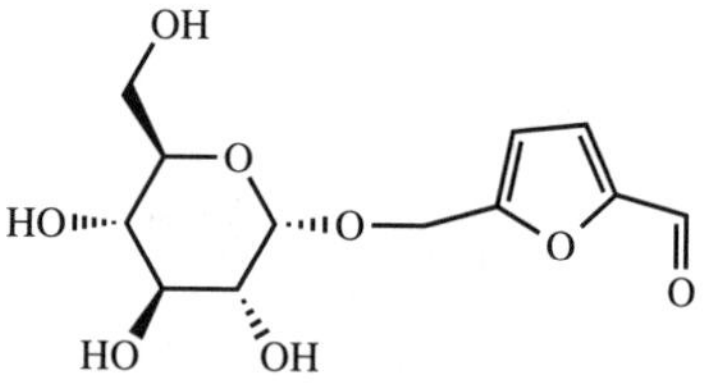

**10.25**
[135213-82-2][23]
Isomaltulose(1)

## BUILDING BLOCK **10.26**

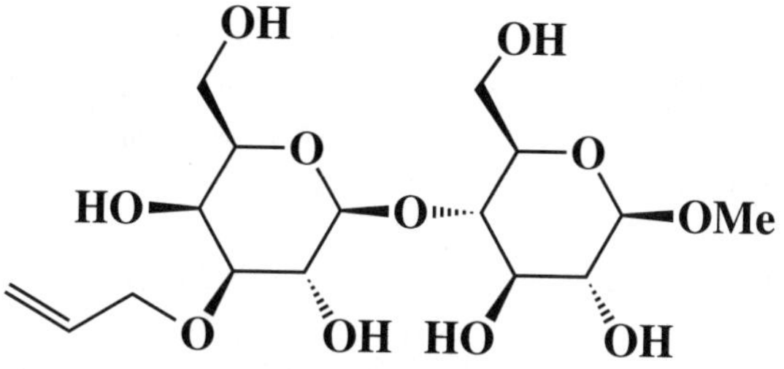

Name: β-D-Glucopyranoside, Methyl 4-*O*-(3-*O*-2-propenyl-β-D-galactopyranosyl-[86733-52-0]

### Synthesis:

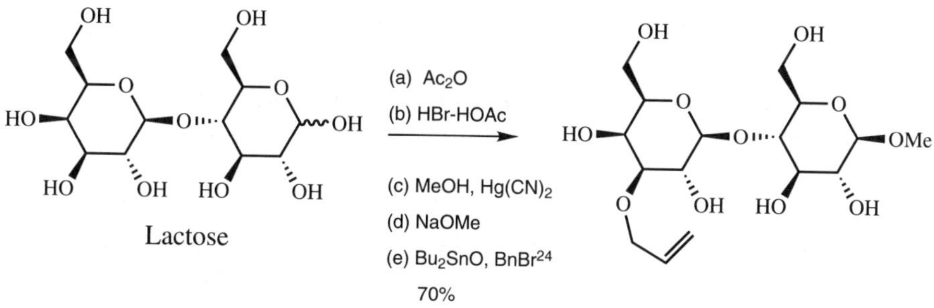

Lactose

(a) $Ac_2O$
(b) HBr-HOAc
(c) MeOH, $Hg(CN)_2$
(d) NaOMe
(e) $Bu_2SnO$, BnBr[24]
70%

### Related Structures:

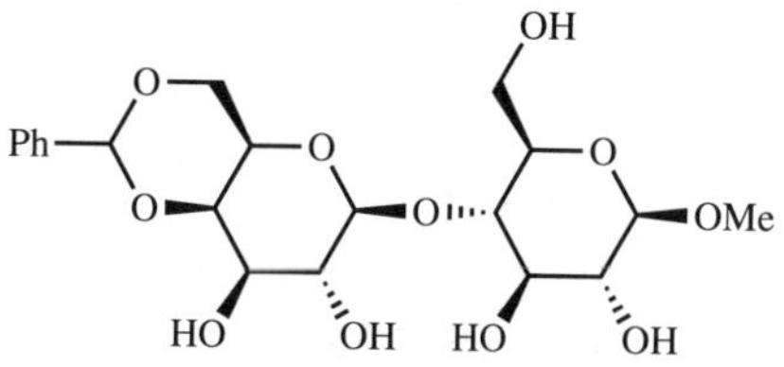

**10.27**
[56865-31-7][25]
Lactose(5)

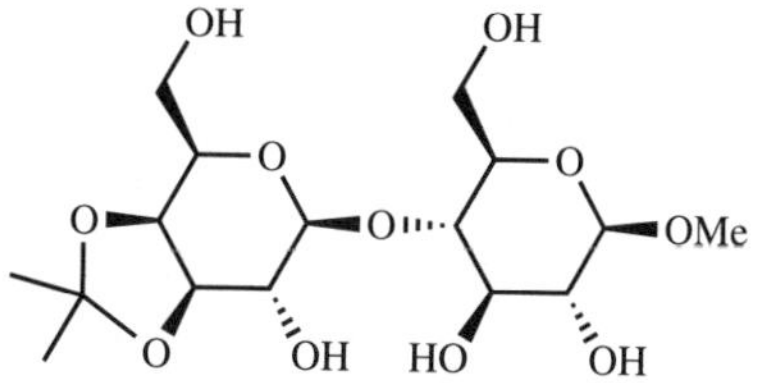

**10.28**
[98169-70-1][26]
Lactose(5)

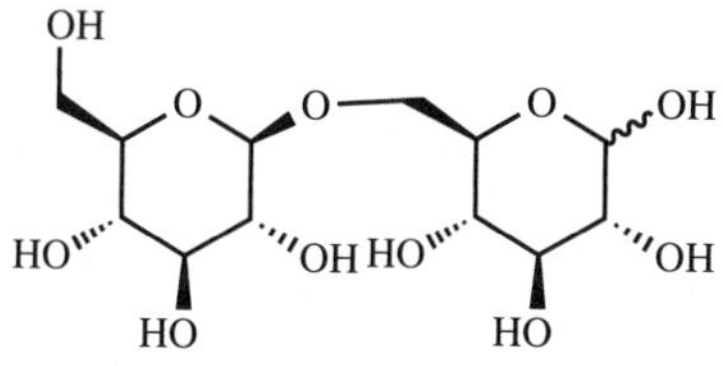

**10.29**
[554-91-6][27]
D-Glucose(1)

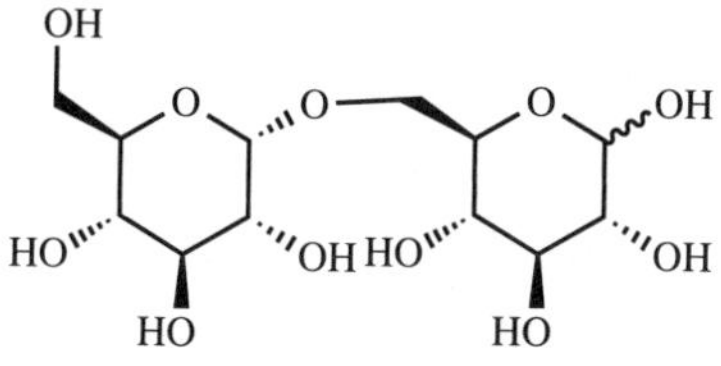

**10.30**
[499-40-1][28]
Maltose(1)

## REFERENCES


1. A. de Raadt and A. E. Stütz. *Tetrahedron Lett.* **1992** 189–192.
2. R. L. Whistler and A. K. Anisuzzaman. *Methods Carbohydr. Chem.* **1980** *8:*227–231.
3. G. G. Makeown, R. S. E. Serenius, and L. D. Hayward. *Can. J. Chem.* **1957** *35:*28–36.
4. M. S. Chowdhary, L. Hough, and A. C. Richardson. *J. Chem. Soc. Perkin Trans. I* **1984** 419–427.
5. F.W. Lichtenthaler. *Modern Synthetic Methods* **1992** *6:*273–376.
6. K. Maurer. *Chem. Ber.* **1930** *63:*25–34.
7. S. Tejima, Y. Okamori. *Chem. Pharm. Bull.* **1972** *20:*2036–2041.
8. F.W. Lichtenthaler, S. Rönninger, and P. Jarglis. *Liebigs Ann.* **1989** 1153–1161.
9. S. Tejima and T. Chiba. *Chem. Pharm. Bull.* **1973** *20:*546–551.
10. L. Asp and B. Lindberg. *Acta Chem. Scand.* **1952** *6:*941–946.
11. K. Koizumi and T. Utamura. *Carbohydr. Res.* **1974** *33:*127–134.
12. S. Tejima and T. Chiba. *Chem. Pharm. Bull.* **1973** *20:*546–551.
13. B. Koeppen. *Carbohydr. Res.* **1970** *13:*193–198.
14. A. M. Rowell and M. S. Feather. *Carbohydr. Res.* **1967** *4:*486–491.
15. W. E. Dick, B. G. Baker, and J. E. Hodge. *Carbohydr. Res.* **1968** *6:*52–62.
16. K. Bock and H. Pedersen. *Acta Chem. Scand.* **1988** *B42:*190–195.
17. F.W. Lichtenthaler, S. Nishiyama, and T. Weiner. *Liebigs Ann.* **1989** 1163–1170.
18. W. N. Haworth, E. L. Hirst, and R. J.W. Reynolds. *J. Chem. Soc.* **1934** 302–303.
19. P. Colson, K. N. Slessor, H. J. Jennings, and I. C. P. Smith. *Can. J. Chem.* **1975** *53:*1030–1037.
20. P. L. Durette, L. Hough, and A. C. Richardson. *Carbohydr. Res.* **1973** *31:*114–119.
21. G. O. Aspinall, T. N. Krishnamurthy, and W. Mitura. *Can. J. Chem.* **1975** *53:*2182–2188.
22. E. Fanton, J. Gelas, and D. Horton. *J. Chem. Soc. Chem. Commun.* **1980** 21–22.
23. F.W. Lichtenthaler, D. Martin, T. A. Weber, and H. M. Schiweck. European Patent Application EP 426,176, **1991**. *Chem. Abstract* **1991** *115:*92826t.
24. J. Alais, A. Maranduba, and A. Veyrieres. *Tetrahedron Lett.* **1983** *24:*2383–2386.
25. R. S. Bhatt, L. Hough, and A. C. Richardson. *Carbohydr. Res.* **1975** *43:*57–67.
26. J. Dahmén, G. Gnosspelins, A.-C. Larsson. T. Lave, G. Noori, K. Pålsson, T. Frejd, and G. Magnusson. *Carbohydr. Res.* **1985** *138:*17–28.
27. I. J. Goldstein and W. J. Whelan. *Methods Carbohydr. Chem.* **1962** *1:*313–315.
28. J. H. Pazur and T. Ando. *Methods Carbohydr. Chem.* **1962** *1:*319–320.

## BUILDING BLOCK **11.01**

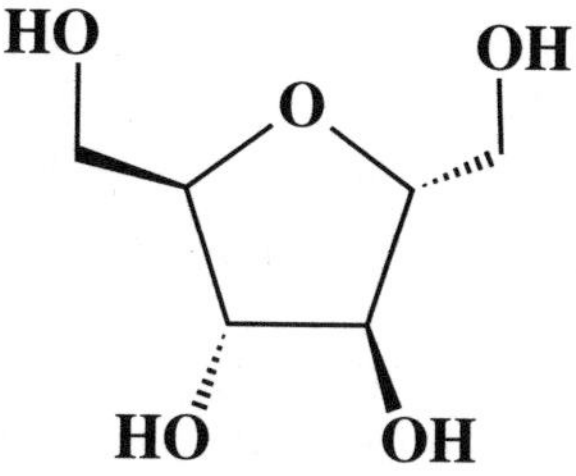

HO OH O HO OH

Name: D-Mannitol, 2,5-Anhydro-[41107-82-8]

### Synthesis:

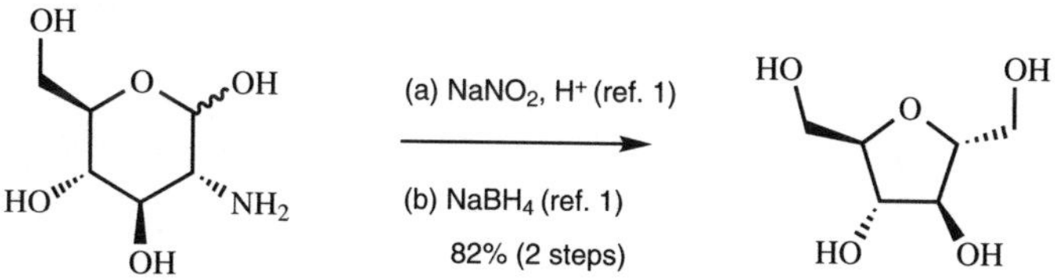

D-Glucosamine

(a) $NaNO_2$, $H^+$ (ref. 1)

(b) $NaBH_4$ (ref. 1)

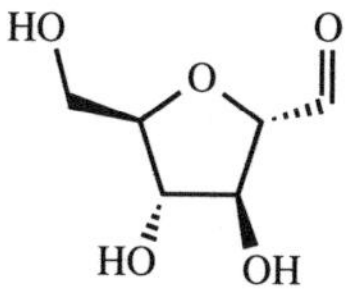

82% (2 steps)

### Related Structures:

| **11.02** | **11.03** | **11.04** | **11.05** |
|---|---|---|---|
| [495-75-0][1] | [4631-21-4][2] | [14218-25-8][2] | [14193-51-2][3] |
| D-Glucosamine(2) | D-Glucose(3) | D-Glucose(3) | D-Mannose(4) |

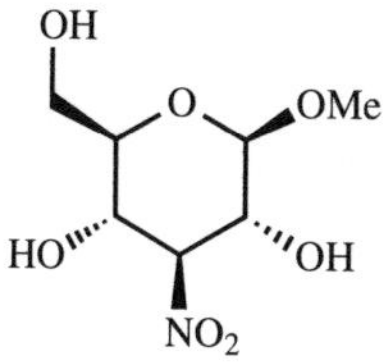

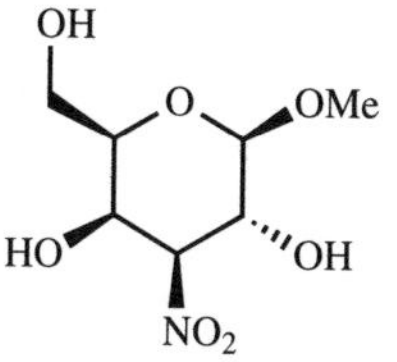

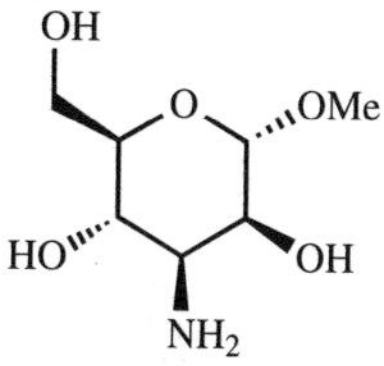

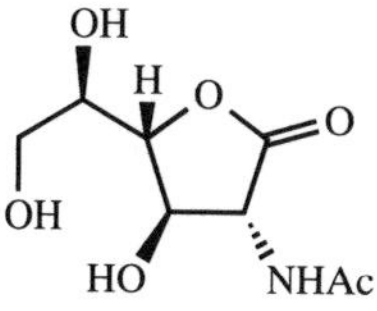

| **11.06** | **11.07** | **11.08** | **11.09** |
|---|---|---|---|
| [28876-38-2][4] | [—][5] | [—][5] | [497-09-6][6] |
| D-Galactose(4) | L-Arabinose(3) | L-Arabinose(3) | L-Sorbose(1) |

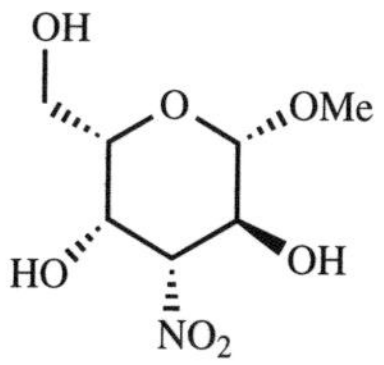

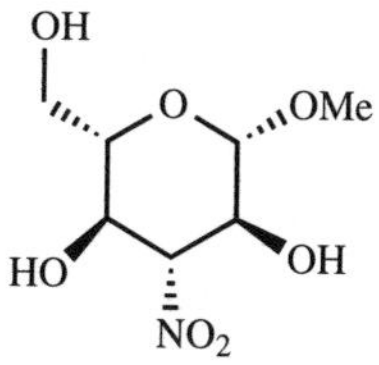

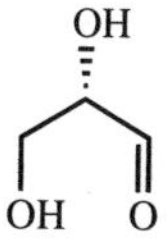

## REFERENCES


1. D. Horton and K. D. Phillips. *Methods Carbohydr. Chem.* **1976** *7:*68–70.
2. H. Baer. *Methods Carbohydr. Chem.* **1972** *6:*245–249.
3. F.W. Lichtenthaler. *Methods Carbohydr. Chem.* **1972** *6:*250–260.
4. V. Zsoldos-Mady, I. Pinter, A. Neszmelyi, A. Messmer, and A. Perczel. *Carbohydr. Res.* **1994** *252:*85–95.
5. H. Baer and F. Kienzle. *Can. J. Chem.* **1965** *43:*3074–3079.
6. A. S. Perlin. *Methods Carbohydr. Chem.* **1962** *1:*61–63.

# STEREOCHEMICAL INDEX

Presented here are stereochemical sequences, and below each sequence we list the numbers of building blocks with that sequence. R is hydrogen, acyl, or alkyl, X is a heteroatom other than oxygen, and C is a carbon substituent. Compounds with a carbon substituent and a hydroxy group (or X) on the same carbon are treated as if the hydroxy function were hydrogen. Compounds in parentheses are disaccharides containing the stereochemical sequence in one of its residues. As backbone the carbon chain with most chiral centers is selected primarily, then the carbon chain in a carbocyclic ring secondly, and finally the longest carbon chain.

**1 Chiral center** OR (D- or L-*glycero*)

| | | | | | |
|---|---|---|---|---|---|
| 2.66 | | | | | |
| 3.79 | 3.80 | | | | |
| 4.41 | 4.55 | 4.59 | 4.102 | | |
| 5.05 | 5.76 | 5.77 | 5.78 | 5.79 | 5.80 |
| 5.81 | | | | | |
| 6.26 | 6.27 | 6.43 | 6.54 | 6.61 | |
| 7.21 | 7.24 | 7.26 | 7.30 | 7.40 | 7.49 |
| 7.50 | 7.51 | 7.52 | 7.54 | 7.64 | |
| 8.24 | 8.32 | | | | |
| 9.01 | 9.05 | | | | |
| (10.05) | (10.24) | | | | |
| 11.09 | | | | | |

**1 Chiral center** (D- or L-*glycero*)

3.38

**1 Chiral center** (D- or L-*glycero*)

8.13 8.15

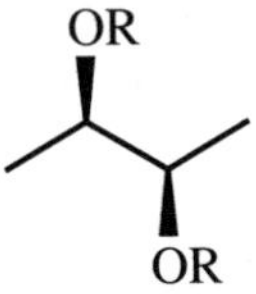

**2 Chiral centers** (D-*threo*)

2.106
4.29 4.39 4.43 4.63
5.36 5.54 5.62 5.63
7.06 7.12 7.16 7.20 7.41 7.47
7.72

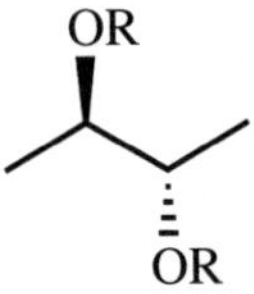

**2 Chiral centers** (D- or L-*erythro*)

1.16
2.10 2.50 2.55 2.65 2.67 2.98
2.101 2.104
3.51
4.01 4.02 4.20 4.22 4.24 4.32
4.34 4.44 4.46 4.48 4.51 4.58
4.62 4.90 4.104
5.53
6.58
7.07 7.08 7.13 7.14 7.19 7.22
7.32 7.38 7.42 7.43 7.46 7.53
7.63
8.07 8.27 8.29 8.31
(10.09) (10.18)

OR
OR

**2 Chiral centers** (L-*threo*)



OR
X

**2 Chiral centers** (D-*threo*)



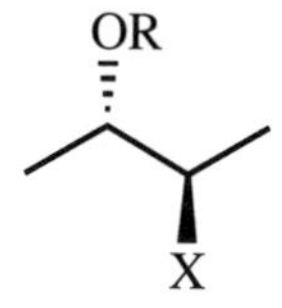


**2 Chiral centers** (D- or L-*erythro*)



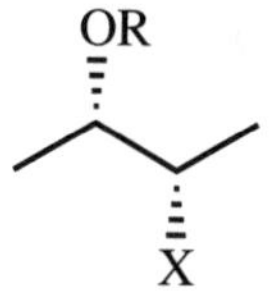


**2 Chiral centers** (L-*threo*)



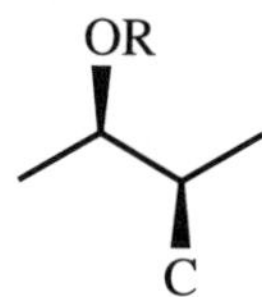


**2 Chiral centers** (D-*threo*)

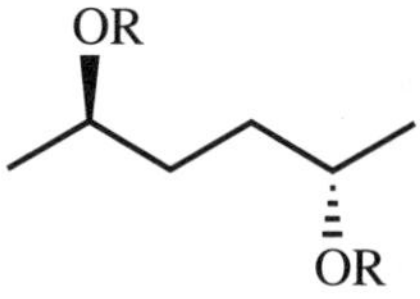

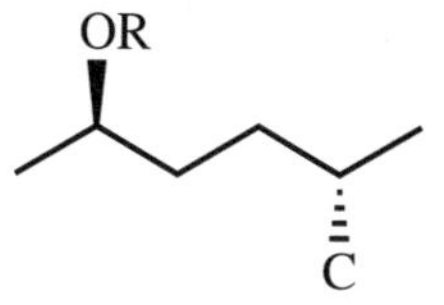
OR
C

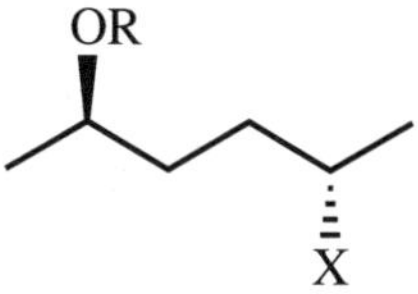

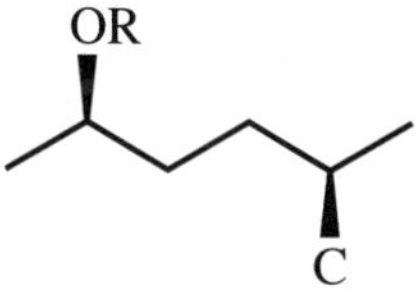

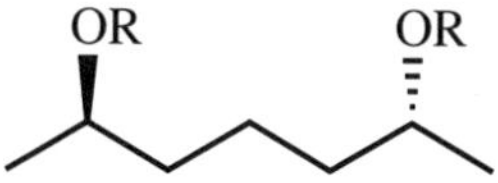

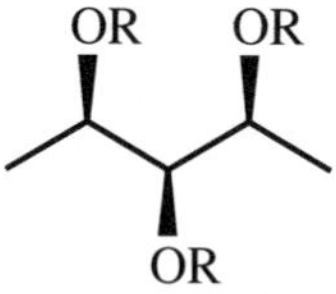

**3 Chiral centers** (D-*arabino* or D-*lyxo*)

| | | | | | |
|---|---|---|---|---|---|
| 1.04 | 1.09 | (1.21) | | | |
| 2.11 | 2.29 | 2.33 | 2.34 | 2.45 | 2.73 |
| 2.75 | 2.93 | | | | |
| 3.10 | 3.11 | 3.12 | 3.20 | 3.25 | 3.35 |
| 3.46 | 3.54 | 3.65 | 3.67 | | |
| 4.04 | 4.05 | 4.27 | 4.37 | 4.45 | 4.49 |
| 4.50 | 4.52 | 4.53 | 4.57 | 4.86 | 4.105 |
| 5.11 | 5.18 | 5.35 | 5.39 | 5.40 | 5.59 |
| 5.60 | 5.61 | | | | |
| 6.07 | 6.32 | 6.34 | 6.48 | | |
| 7.01 | 7.02 | 7.04 | 7.05 | 7.10 | 7.11 |
| 7.15 | 7.17 | 7.37 | 7.39 | 7.65 | 7.66 |
| 7.69 | 7.71 | | | | |
| 8.19 | 8.20 | | | | |
| (10.01) | (10.02) | (10.03) | (10.04) | (10.05) | (10.06) |
| (10.07) | (10.19) | | | | |

**3 Chiral centers** (D- or L-*ribo*)

| | | | | | |
|---|---|---|---|---|---|
| 1.06 | | | | | |
| 2.13 | 2.18 | 2.32 | 2.62 | 2.68 | 2.72 |
| 2.94 | 2.100 | | | | |
| 3.61 | | | | | |
| 4.08 | 4.33 | 4.35 | 4.75 | 4.82 | 4.83 |
| 4.95 | | | | | |
| 5.41 | 5.42 | 5.44 | | | |
| 6.23 | | | | | |
| 7.28 | 7.68 | | | | |

**3 Chiral centers** (L-*arabino* or L-*lyxo*)

| | |
|---|---|
| 1.05 | |
| 2.12 | 2.74 |
| 3.04 | 3.36 |
| 4.72 | 4.76 |
| 5.17 | 5.38 |
| 7.03 | |
| 8.28 | |

**3 Chiral centers** (D- or L-*xylo*)

4.38 4.40 4.98

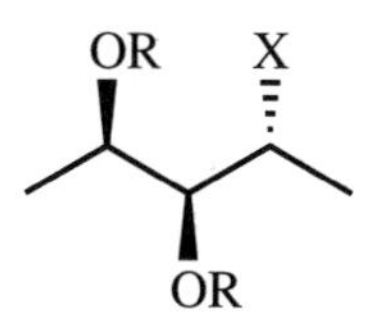


**3 Chiral centers** (D-*arabino* or D-*lyxo*)

3.45
4.36
7.31

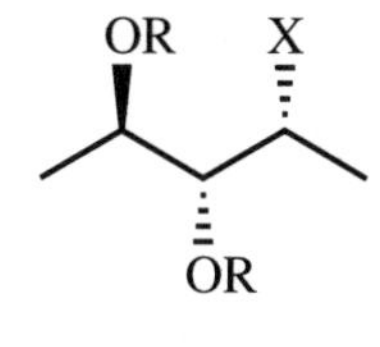


**3 Chiral centers** (D-*arabino* or D-*lyxo*)

4.31 4.42

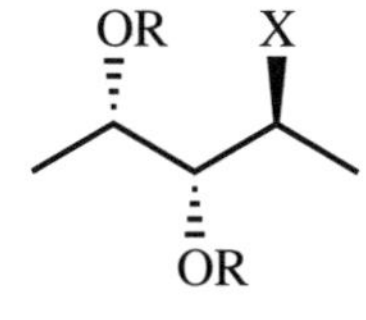


**3 Chiral centers** (L-*arabino* or L-*lyxo*)

3.44

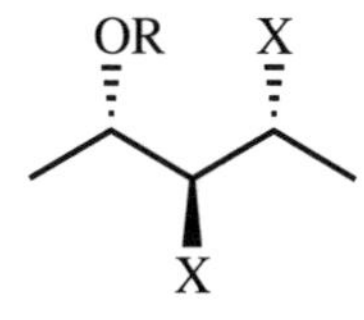


**3 Chiral centers** (D- or L-*ribo*)

3.39

**3 Chiral centers** (D- or L-*ribo*)

2.88
6.53

OR C

OR

**3 Chiral centers** (D-*arabino* or D-*lyxo*)

2.87

8.01 8.02 8.03 8.04 8.33

OR C

OR

**3 Chiral centers** (L-*arabino* or L-*lyxo*)

7.58

OR C

OR

**3 Chiral centers** (L-*arabino* or L-*lyxo*)

2.108

OR C

OR

**3 Chiral centers** (D- or L-*ribo*)

2.107

OR OR

C

**3 Chiral centers** (D-*arabino* or D-*lyxo*)

3.62

OR OR

C

**3 Chiral centers** (L-*arabino* or L-*lyxo*)

3.63

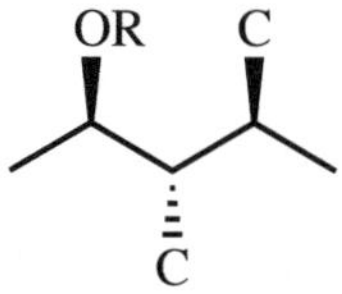

OR C

C

**3 Chiral centers** (D- or L-*ribo*)



OR

OR OR

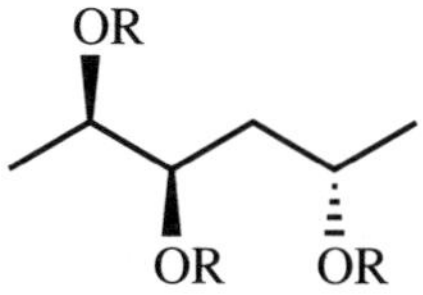

**3 Chiral centers** (D- or L-*xylo*)



OR

OR OR

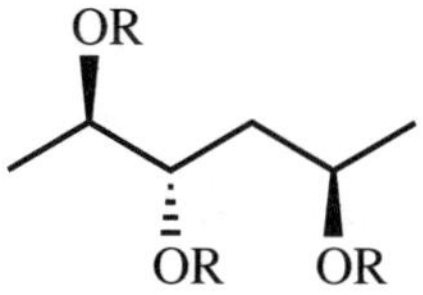

**3 Chiral centers** (D-*arabino* or D-*lyxo*)



OR

OR OR

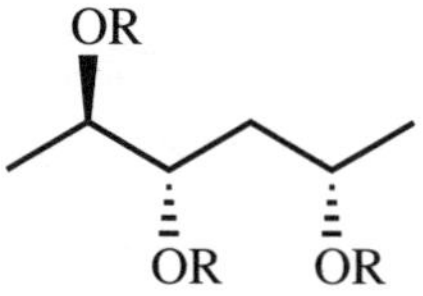

**3 Chiral centers** (D- or L-*ribo*)



OR

OR OR

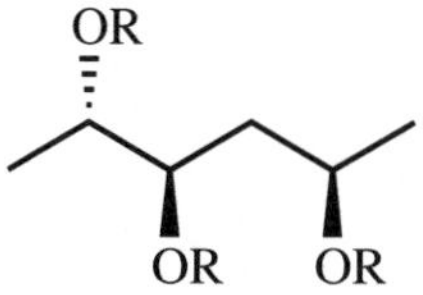

**3 Chiral centers** (D- or L-*ribo*)



OR

OR OR

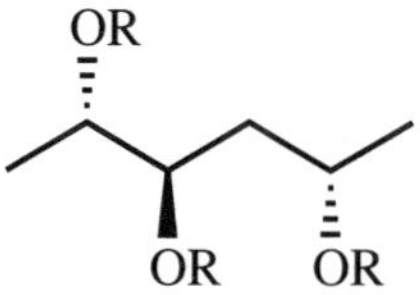

**3 Chiral centers** (L-*arabino* or L-*lyxo*)

**3 Chiral centers** (D- or L-*xylo*)



**3 Chiral centers** (D-*aribino* or D-*lyxo*)



**3 Chiral centers** (D-*arabino* or D-*lyxo*)



**3 Chiral centers** (D- or L-*ribo*)



**3 Chiral centers** (D-*arabino* or D-*lyxo*)



**4 Chiral centers** (D-*ido*)

OR OR

OR OR

**4 Chiral centers** (L-*gluco* or D-*gulo*)

| | | | | | |
|---|---|---|---|---|---|
| 2.85 | 2.91 | 2.92 | | | |
| 3.73 | | | | | |
| 4.78 | 4.81 | | | | |
| 5.74 | | | | | |

OR OR

OR OR

**4 Chiral centers** (D-*altro* or D-*talo*)

| | | | | | |
|---|---|---|---|---|---|
| 2.53 | 2.76 | | | | |
| 3.74 | | | | | |
| 6.08 | 6.16 | 6.17 | 6.52 | | |
| 9.04 | | | | | |

OR OR

OR OR

**4 Chiral centers** (D- or L-*galacto*)

| | | | | | |
|---|---|---|---|---|---|
| 1.02 | (1.24) | | | | |
| 2.15 | 2.16 | 2.19 | 2.21 | 2.23 | 2.49 |
| 2.86 | | | | | |
| 3.08 | 3.14 | 3.22 | 3.32 | 3.68 | 3.70 |
| 3.75 | 3.78 | | | | |
| 4.09 | 4.10 | 4.84 | | | |
| 5.47 | 5.52 | 5.69 | 5.71 | | |
| 6.02 | 6.04 | 6.12 | 6.20 | 6.50 | |
| 7.62 | | | | | |
| (10.09) | (10.10) | (10.26) | (10.27) | (10.28) | |

OR OR

OR OR

**4 Chiral centers** (D- or L-*allo*)

| | | | | | |
|---|---|---|---|---|---|
| 2.02 | 2.25 | 2.46 | 2.64 | | |
| 6.18 | | | | | |

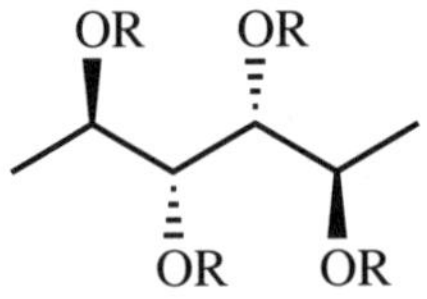

**4 Chiral centers** (D-*manno*)

| | | | | | |
|---|---|---|---|---|---|
| 1.03 | 1.20 | | | | |
| 2.14 | 2.20 | 2.22 | 2.28 | 2.47 | 2.54 |
| 2.102 | 2.103 | | | | |
| 3.09 | 3.13 | 3.18 | 3.34 | 3.69 | 3.76 |
| 4.06 | 4.18 | 4.88 | 4.93 | | |
| 5.03 | 5.06 | 5.07 | 5.10 | 5.15 | 5.22 |
| 5.23 | 5.27 | 5.48 | 5.58 | 5.72 | |
| 6.03 | 6.05 | 6.06 | 6.10 | 6.15 | 6.21 |
| 6.51 | | | | | |
| 11.01 | 11.02 | | | | |

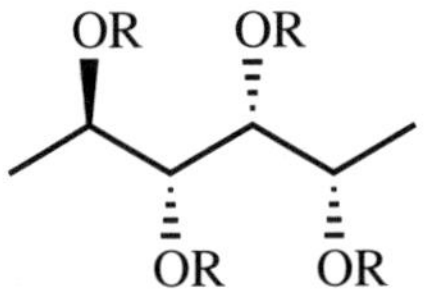

**4 Chiral centers** (D-*gluco* or L-*gulo*)

| | | | | | |
|---|---|---|---|---|---|
| 1.01 | 1.11 | 1.12 | 1.14 | 1.17 | 1.19 |
| 1.21 | (1.23) | (1.24) | (1.25) | (1.26) | |
| 2.01 | 2.03 | 2.26 | 2.27 | 2.30 | 2.31 |
| 2.48 | 2.51 | 2.56 | 2.58 | 2.60 | 2.61 |
| 2.71 | 2.80 | 2.83 | 2.89 | | |
| 3.01 | 3.03 | 3.06 | 3.07 | 3.19 | 3.23 |
| 3.26 | 3.28 | 3.29 | 3.30 | 3.31 | 3.33 |
| 3.35 | 3.47 | 3.52 | 3.64 | 3.66 | 3.81 |
| 4.07 | 4.30 | 4.73 | 4.77 | 4.79 | 4.80 |
| 4.85 | 4.87 | 4.89 | 4.94 | | |
| 5.01 | 5.02 | 5.09 | 5.16 | 5.19 | 5.21 |
| 5.24 | 5.29 | 5.30 | 5.31 | 5.37 | 5.46 |
| 5.49 | 5.57 | 5.73 | | | |
| 6.01 | 6.11 | 6.19 | | | |
| 7.61 | | | | | |
| (10.01) | (10.02) | (10.03) | (10.04) | (10.06) | (10.07) |
| (10.08) | (10.10) | (10.11) | (10.12) | (10.13) | (10.14) |
| (10.15) | (10.16) | (10.17) | (10.18) | (10.19) | (10.20) |
| (10.22) | (10.23) | (10.24) | (10.25) | (10.26) | (10.27) |
| (10.28) | (10.29) | (10.30) | | | |

OR X / OR OR

**4 Chiral centers** (D- or L-*allo*)
3.40
(10.21)

OR X / OR OR

**4 Chiral centers** (D-*manno*)
11.05

OR X / OR OR

**4 Chiral centers** (D-*gluco* or L-*gulo*)
11.03

OR X / OR OR

**4 Chiral centers** (L-*gluco* or D-*gulo*)
11.08

OR X / OR OR

**4 Chiral centers** (L-*altro* or L-*talo*)
2.59

OR X / OR OR

**4 Chiral centers** (D- or L-*galacto*)
3.15 3.37 3.75
(10.21)
11.07

OR X / OR X

**4 Chiral centers** (D-*altro* or L-*talo*)
2.81

**4 Chiral centers** (D- or L-*allo*)
3.42

**4 Chiral centers** (D- or L-*galacto*)
7.57

**4 Chiral centers** (D-*altro* or D-*talo*)
2.69

**4 Chiral centers** (D-*gluco* or L-*gulo*)
6.14 6.45
7.59
8.21 8.34

**4 Chiral centers** (D-*altro* or D-*talo*)
2.70

**4 Chiral centers** (D- or L-*allo*)
2.05 2.08 2.38 2.39

**4 Chiral centers** (D-*gluco* or L-*gulo*)
2.42 2.43

OR C

OR OR

**4 Chiral centers**
2.95

(D- or L-*galacto*)

OR C

OR OR

**4 Chiral centers**
2.96
6.13

(D-*gluco* or L-*gulo*)

OR C

OR C

**4 Chiral centers**
7.33

(D- or L-*allo*)

OR C

C OR

**4 Chiral centers**
6.42

(D-*manno*)

OR OR OR

OR

**4 Chiral centers**
4.65 4.66

(D-*gluco* or L-*gulo*)

OR OR

OR OR

**4 Chiral centers**
7.44

(D- or L-*allo*)

C C OR

OR

**4 Chiral centers**
8.09

(D-*altro* or D-*talo*)

**5 Chiral centers** (D-*glycero*-D-*gulo* or L-*glycero*-L-*gulo*)

1.13
4.13 4.91

**5 Chiral centers** (L-*glycero*-D-*gluco* or L-*glycero*-D-*galacto*)

6.22

**5 Chiral centers** (D-*glycero*-D-*ido* or D-*glycero*-L-*gulo*)

4.12

**5 Chiral centers** (D-*glycero*-L-*galacto* or D-*glycero*-L-*gluco*)

6.37 6.38

**5 Chiral centers** (D-*glycero*-D-*gluco* or L-*glycero*-D-*altro*)

6.39

**5 Chiral centers** (L-*glycero*-D-*galacto or* L-*glycero*-L-*gluco*)

6.40

OR OR OR

OR X

**5 Chiral centers**

(D-*glycero*-D-*ido* or D-*glycero*-L-*gulo*)

8.36

OR OR OR

OR OR X

**6 Chiral centers**

2.105

(L-*threo*-D-*galacto*)

OR OR C OR

OR OR X

**7 Chiral centers**

(D-*lyxo*-D-*gulo* or L-*arabino*-L-*galacto*)

7.55

# INDEX OF PARTIALLY PROTECTED COMPOUNDS

Partially protected compounds are listed with the indicated unprotected OH groups. Only the most common sugars are listed. While stereochemistry is depicted as in the stereochemical index, general carbohydrate numbering is used.

| **Compound** | **Stereochemistry** | **Unprotected OH Groups** | | | |
|---|---|---|---|---|---|
| *Galactose* | 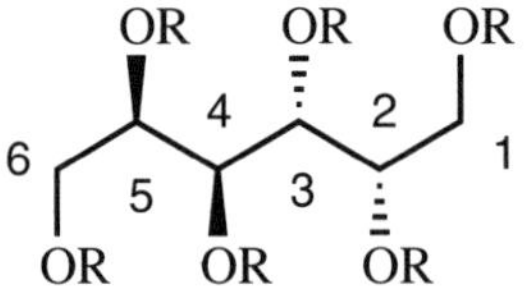 (D-/L-*galacto*) | *2* | *3* | 4 | 6 |
| | | 2.86 | 2.23 | 3.08 | 2.15 |
| | | 6.04 | | | 3.78 |
| | | *2,3* | *2,4* | *2,6* | *3,4* |
| | | 2.49 | 3.14 | 2.16 | 3.68 |
| | | 4.84 | | | |
| | | *1,2,3* | *2,3,4* | *2,3,6* | *2,4,6* |
| | | 2.19 | 5.52 | 5.47 | 3.71 |
| | | | 6.02 | | |
| | | *3,4,6* | | | |
| | | 2.21 | | | |
| *Mannose* | 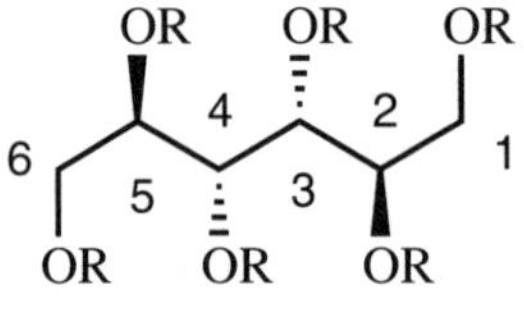 (D-*manno*) | *1* | 2 | *3* | *4* |
| | | 2.22 | 2.102 | 2.103 | 3.09 |
| | | | 3.18 | | 6.05 |
| | | | 5.48 | | |
| | | | 6.06 | | |
| | | *6* | *1,6* | *2,4* | *2,5* |
| | | 3.76 | 5.14 | 3.13 | 5.13 |
| | | | | 3.69 | 5.58 |

| Compound | Stereochemistry | Unprotected OH Groups | | | |
|---|---|---|---|---|---|
| | | *3,4* | *3,5* | *5,6* | *1,2,3* |
| | | 5.03 | [4.18] | 2.14 | 2.28 |
| | | 5.07 | | 5.15 | |
| | | [6.15] | | | |
| | | *2,3,4* | *1,2,5,6* | *2,3,4,5* | |
| | | 6.03 | 5.10 | 5.06 | |
| *Glucose* | 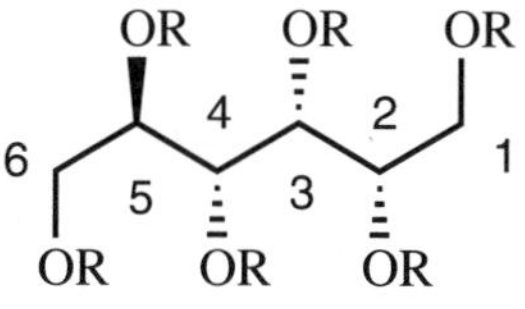 (D-*gluco*-/L-*gulo*) | *1* | *2* | *3* | *4* |
| | | 3.66 | 3.19 | 2.01 | 2.61 |
| | | 5.31 | 4.85 | [2.71] | 5.49 |
| | | | | 2.89 | |
| | | | | 3.64 | |
| | | | | [6.11] | |
| | | *5* | *6* | *1,2* | *1,5* |
| | | 2.83 | 2.58 | 2.80 | 4.79 |
| | | 4.80 | 2.60 | 4.73 | |
| | | | 3.06 | | |
| | | *2,3* | *3,4* | *5,6* | *1,2,3* |
| | | 2.26 | 3.01 | 5.02 | 2.31 |
| | | 2.48 | [3.07] | | 2.51 |
| | | | 5.09 | | |
| | | *2,3,4* | *3,5,6* | *1,2,4,6* | *1,2,5,6* |
| | | 3.03 | 2.02 | 5.57 | 5.16 |
| | | 6.01 | 5.46 | | |
| | | *1,3,5,6* | | | |
| | | 5.57 | | | |
| *Rhamnose* | 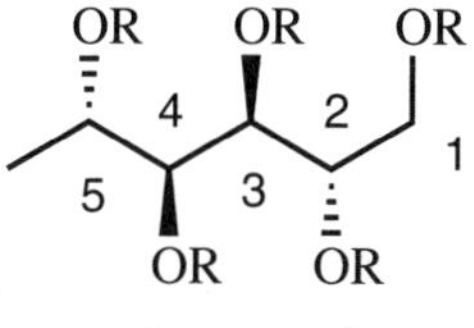 (L-*manno*) | *3* | *4* | *1,5* | *2,4* |
| | | 3.77 | 2.17 | 2.24 | 3.70 |
| | | | 3.72 | | |
| | | *3,5* | | | |
| | | [4.15] | | | |
| *Fructose* | 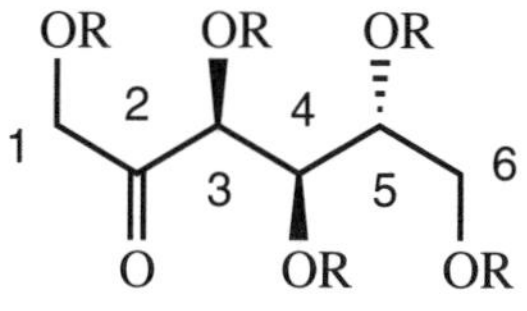 (D-*arabino*/D-*lyxo*) | *1* | *2* | *3* | *1,2,6* |
| | | 2.29 | 3.10 | 2.33 | 5.59 |
| | | | 3.11 | | |
| | | | 3.12 | | |
| | | | 3.67 | | |
| | | *3,4,5* | | | |
| | | 2.34 | | | |

| Compound | Stereochemistry | Unprotected OH Groups | | | |
|---|---|---|---|---|---|
| *Sorbose* | OR OR OR<br>2 4<br>1<br>3 5 6<br>O OR OR<br>(D/L-*xylo*) | *1*<br>2.57 | *2,4*<br>[2.78] | *1,4,6*<br>2.35 | *3,4,5*<br>2.36 |
| *Xylose* | OR OR<br>3<br>5 1<br>4 2<br>OR OR OR<br>(D/L-*xylo*) | *2*<br>2.52 | *2,4*<br>3.60 | *3,5*<br>2.04 | *1,3,5*<br>5.08 |
| D-*Arabinose* | OR OR<br>3<br>1 5<br>2 4<br>OR OR OR<br>(D-*arabino*/D-*lyxo*) | *1*<br>2.75<br>3.65<br>4.86<br>5.18<br>*2,4,5*<br>5.11 | *2*<br>2.11 | *3*<br>[4.27] | *4,5*<br>5.40 |
| *Ribose* | OR OR<br>3<br>5 1<br>4 2<br>OR OR OR<br>(D/L-*ribo*) | *2*<br>4.83<br>*3,5*<br>5.41 | *5*<br>2.13<br>4.82 | *1,5*<br>2.18<br>2.62 | *2,4*<br>3.61 |
| L-*Arabinose* | OR OR<br>3<br>5 1<br>4 2<br>OR OR OR<br>(L-*arabino*/L-*lyxo*) | *1*<br>5.17<br>5.38 | *2*<br>2.12 | *3*<br>[2.74] | *4*<br>3.04 |

# CARBOCYCLIC INDEX

Presented here is an index of compounds containing carbocyclic rings.

# INDEX OF COMPOUNDS WITH BRANCHED CARBON CHAIN

| Branch Points | Other Chiral Centers | Compounds | |
|---|---|---|---|
| 1 | 0 | 8.13 | 8.15 |
| | | 2.107 | 2.108 |
| | | 4.60 | 4.101 |
| | | 4.102 | 7.23 |
| | | 7.24 | 7.25 |
| | | 7.27 | 7.35 |
| | | 7.36 | 8.05 |
| 1 | 1 | 8.06 | 8.10 |
| | | 8.11 | 8.12 |
| | | 8.14 | 8.16 |
| | | 8.17 | 8.18 |
| | | 8.22 | 8.23 |
| | | 8.24 | 8.25 |
| | | 8.26 | 8.32 |
| | | 8.35 | |
| | | 2.87 | 2.88 |
| | | 2.101 | 3.62 |
| | | 3.63 | 7.43 |
| 1 | 2 | 7.58 | 8.01 |
| | | 8.02 | 8.03 |
| | | 8.07 | 8.27 |
| | | 8.29 | 8.30 |
| | | 8.31 | 8.33 |

| Branch Points | Other Chiral Centers | Compounds | |
|---|---|---|---|
| | | 2.05 | 2.06 |
| | | 2.08 | 2.38 |
| | | 2.39 | 2.42 |
| | | 2.43 | 2.69 |
| 1 | 3 | 2.70 | 2.95 |
| | | 2.96 | 2.99 |
| | | 2.100 | 5.61 |
| | | 6.13 | 6.14 |
| | | 6.45 | 7.57 |
| | | 7.59 | 8.21 |
| | | 8.34 | |
| 1 | 4 | 6.37 | 6.38 |
| | | 6.39 | 6.40 |
| 1 | 6 | 7.55 | |
| | | 4.100 | 6.41 |
| 2 | 1 | 6.44 | 7.34 |
| | | 7.56 | |
| 2 | 2 | 6.42 | 7.33 |
| | | 8.08 | 8.09 |

# SUBJECT INDEX

NORTHERN MICHIGAN UNIVERSITY
3 1854 005 404 119